DAVID BIRD trained as an analytical chemist and entered the food manufacturing business as an analyst working with baby foods, mustard and fruit squashes. He moved into the wine trade in 1973 almost by chance, but in reality because a passion for wine was already developing. 1981 was his vintage year, becoming a Master of Wine, a Chartered Chemist and father to his first son. He specialises in quality assurance techniques, such as ISO 9000 and HACCP, and has been involved with wine activities and education in France, Italy, Spain, Portugal, Hungary, Denmark, Finland, Sweden, Norway, the Netherlands, Ukraine, Moldova, Russia, Algeria, Australia, Scotland, Ireland and England.

He plays the organ and is Musical Director of the Cantate choir. His garden is open to the public once a year for charity under the National Gardens Scheme of the UK.

.

UNDERSTANDING WINE TECHNOLOGY

A BOOK FOR THE NON-SCIENTIST
THAT EXPLAINS THE SCIENCE OF WINEMAKING

DAVID BIRD

Chartered Chemist and Master of Wine

DBQA Publishing, Great Britain

In association with The Wine Appreciation Guild, USA

Text copyright © 2000, 2005, 2010 David Bird

First published in Great Britain in 2000 by
DBQA Publishing
www.dbqa.com

First published in the USA in 2005 by
Wine Appreciation Guild (USA)
an imprint of
Board and Bench Publishing
www.boardandbench.com

UK edition: ISBN 978-0-9535802-2-4
US edition: ISBN 978-1-934259-60-3

Designed by David Bird
Photography and drawings by David Bird
US cover typeset by Diane Spencer Hume

Although all reasonable care has been taken in the preparation of this book,
neither the author nor the publisher can accept liability for any consequences
arising from the information contained herein or from use thereof.

Printed by TC Transcontinental Printing
150, 181st Street, Beauceville, Quebec G5X 3P3, Canada

Wine, the most delightful of drinks, whether we owe it to Noah, who planted the vine, or to Bacchus, who pressed the juice from the grape, goes back to the childhood of the world.

BRILLAT-SAVARIN

Contents

Chapter 1 **THE GIFT OF NATURE**..1
The origins of wine..1
The natural cycle..2
Enzymes in nature..4
Wine and health..5
    Alcohol..6
    Flavonoids..6
    Resveratrol..7
    Potassium..7
    Histamine..7
The modern paradox..8

Chapter 2 **IN THE VINEYARD**..9
The vine..9
Phylloxera & grafting..11
Climate 12
Training & pruning..13
Soil & water..14
Irrigation..16
Green harvest..16
Terroir..17
Vineyard systems..18
    Viticulture Raisonée (La Lutte Raisonée)..18
    Organic viticulture..19
    Biodynamic viticulture..19

Chapter 3 **INSIDE THE GRAPE**..21
Sugars..21
Acids..23
Mineral salts..24
Phenolic compounds..25
    Tannins..25
    Anthocyanins..26
Flavour components..27
Proteins & colloids..28
Véraison & maturity..29

viii

Chapter 4 **THE ROLE OF OXYGEN**..**31**
   Old-style winemaking..32
   Anaerobic winemaking...32
   Antioxidants 33
   Inert gases..33
      Carbon dioxide...34
      Nitrogen...35
      Argon..36
   Dissolved oxygen..36
   Sparging..37
   The positive role of oxygen...................................37

Chapter 5 **PRODUCING THE MUST**............................**39**
   Harvesting the grapes...40
      Picking by hand..40
      Machine harvesting...41
   Transport to the winery.......................................43
   Sorting...45
   De-stemming 45
   Crushing the grapes...47
   Draining the juice...48
   Pressing the berries...49
      The basket press...49
      Horizontal screw press.....................................50
      Pneumatic press..53
      Tank press..53
      Continuous screw press.....................................55

Chapter 6 **ADJUSTING THE MUST**............................**57**
   Sulphur dioxide...59
   Clarification (white and pink wines)..........................60
      Settling...60
      Centrifuging..60
      Flotation..61
   Hyperoxidation..62
   Acidification...63
   Deacidification..64
      Acidex..65
   Enrichment..65
   Must concentration..67
      Vacuum distillation...67
      Cryo-extraction (cryo-concentration).......................68

Reverse osmosis..68
Nutrients..69
Other treatments..70
    Bentonite..70
    Activated charcoal..70
    Tannin..70

Chapter 7 **FERMENTATIONS**..**71**
Yeasts..72
The action of yeasts...74
Natural fermentation..76
Cultured yeasts..77
Control of temperature...78
Monitoring the fermentation...80
Stopping the fermentation..81
A 'stuck' fermentation..83
Naturally sweet wines..84
The malo-lactic fermentation..84

Chapter 8 **RED & PINK WINE PRODUCTION**....................**87**
Fermentation vessels..88
Maceration...89
Traditional process...90
Submerged cap process..92
Pumping-over systems..92
Délestage (Rack and return)..94
Autovinifier 94
Rotary fermenters...96
Thermo-vinification..96
Flash détente...97
Carbonic maceration (Macération carbonique)........................97
Variants on carbonic maceration...100
    Whole bunch fermentation..100
    Whole berry fermentation..101
Pink wines..101
    Short maceration...102
    Saignée 102
    Vin d'une nuit 102
    Double pasta 102

Chapter 9 **WHITE WINE PRODUCTION**..............................**103**
   Cool fermentation...103
   Skin contact (macération pelliculaire)......................104
   Tank vs. barrel..105
   Sur lie..106
   Bâtonnage...106
   Prevention of oxidation...107
   Sweet wines..108
   Carafe wines...108
   German wines..109
   Sauternes...110
   Tokaji Aszú...111

Chapter 10 **SPARKLING & FORTIFIED PROCESSES**......................**113**
   Sparkling wines...113
      Traditional method...113
      Transfer method...116
      Tank method (Cuve Close or Charmat)................116
      Carbonation ("Pompe bicyclette")......................117
      The Asti method 117
   Fortified wines (liqueur wines)................................117
      Vins doux naturelles (VDN).............................118
      Port 118
      Port styles...120
      Sherry 121
      Madeira...124
      Marsala...124

Chapter 11 **WOOD & MATURATION**.....................................**125**
   Type of wood..126
   Oak...126
   Size of vessel..128
   Seasoning and toasting...129
   Fermentation in barrel...130
   Maturation in wood..132
   Putting the wood in the wine...................................132
   Micro-oxygenation...133

Chapter 12 **PRINCIPAL COMPONENTS OF WINE**......................**135**
   Alcohols..135
   Acids...137
   Esters..138

Residual sugars...140
Glycerol 141
Aldehydes and ketones...142

Chapter 13 **CLARIFICATION & FINING**..........................**143**
Is treatment necessary?..143
Racking..144
Protection from oxidation...145
Blending...145
Fining...146
Fining agents..149
    Ox blood..149
    Egg white 150
    Albumin...150
    Gelatine...150
    Isinglass (ichthyocol or Col de poisson)....................151
    Casein..151
    Tannin..151
    Bentonite...151
    Silica sol (Kieselsol)..152
    Polyvinylpolypyrrolidone (PVPP)..............................152
    Activated charcoal...152
    Allergens...152
Blue fining..153
Calcium phytate..155
PVI/PVP copolymers...155
Chitin-glucan complex and chitosan...................................156

Chapter 14 **TARTRATE STABILISATION**............................**157**
Natural and harmless?...157
Cold stabilisation..158
Contact process 159
Ion exchange..160
Electrodialysis..161
Metatartaric acid..162
Carboxymethylcellulose (Cellulose gums)..........................163
Mannoproteins...163
Minimum intervention..164

Chapter 15 **ADDITIVES**..**165**
Sulphur dioxide...165
    Free and total sulphur dioxide...................................171

Molecular sulphur dioxide..173
Ascorbic acid...174
Sorbic acid...175
Metatartaric acid..176
Citric acid..176
Copper sulphate or silver chloride................................176
Acacia (Gum arabic)...177
Enzymes..178
    Pectinolytic enzymes..178
    Betaglucanase...179
    Lysozyme..179
    Laccase...180
    Tyrosinase...180

Chapter 16 **FILTRATION**...**181**
Principles of filtration..182
Depth filters..183
    Kieselguhr filtration (earth filtration)......................183
    Sheet filtration (plate & frame or pad filtration)........186
Surface filters...189
    Membrane filtration (Cartridge filtration) 189
    Cross-flow filtration (tangential filtration)...............192
Ultrafiltration..194

Chapter 17 **PACKAGING MATERIALS**............................**195**
Containers..195
    Glass bottles...195
    Measuring container bottles (MCBs).........................196
    Plastic bottles...197
    Aluminium cans...198
    Bag-in-box (BIB)...198
    Cardboard 'bricks'...201
Closures..202
    Natural cork..202
    Technical corks..203
    Synthetic closures..205
    Aluminium screwcaps..206
    Glass stoppers...208
    Other closures...208

Capsules..209
    Lead foil...209
    Pure tin..209
    Tin-lead...209
    Aluminium...210
    PVC...210
    Polylaminated..210

Chapter 18 **STORAGE & BOTTLING**..................................**211**
    Storage without change...211
    The final sweetening...213
    Shipping in bulk...214
    Bottling processes..214
    Traditional bottling..214
    'Sterile' bottling..215
    Principles of modern bottling..216
    Dimethyldicarbonate (DMDC)..217
    Modern bottling techniques...217
        Bottle rinsing..218
        Thermotic, or Hot Bottling...219
        Tunnel pasteurisation..220
        Flash pasteurisation...222
        Cold sterile filtration...223
    Maturation in bottle...225

Chapter 19 **QUALITY CONTROL & ANALYSIS**..................**227**
    Quality plan..228
    Records and traceability..228
    Laboratory analyses...229
        Density...231
        Alcoholic strength...232
        Total dry extract (TDE)...235
        Total acidity...235
        pH...237
        Volatile acidity...238
        Residual sugars...239
        Tartrate stability tests...240
        Protein stability tests..241
    Permitted additives..241
        Sulphur dioxide..241

Other additives...243
Contaminants..244
    Dissolved oxygen...244
    Iron and copper..244
    Sodium...246
Advanced methods of analysis..246
Microbiological analysis..247

Chapter 20 **WINE FAULTS**...**249**
Oxidation..249
Reductive taint..250
Beyond shelf-life..251
Tartrate crystals...252
Foreign bodies..254
Musty taint...254
Volatile acidity..255
Second fermentation...255
Iron casse...256
Copper casse..256
Mousiness...257
Brett 257
Geranium taint..258

Chapter 21 **THE TASTE TEST**..**259**
Preparations for tasting..259
    Temperature...259
    Decanting..260
Tasting (or drinking) glasses...260
Styles of tasting..262
    Tasting in front of the label.......................................262
    Comparative tasting..262
    Blind tasting...262
Writing a tasting note...263
Tasting the wine...265
    Appearance...266
    Nose..266
    Palate...267
    Conclusion..268
Drinking - A few personal tips:.......................................268

Chapter 22 **QUALITY ASSURANCE**......................................**269**
    Hazard analysis and critical control points(HACCP)..........................270
        Principles of the HACCP system...271
        The process..271
        The application to a winery...274
        Interpretation..278
    ISO 9001:2008..278
    ISO 14000:2004..280
    ISO 22000:2005..280
    Supplier audits..281
    The BRC Global Food Standard..282
    Business Excellence Model..283

Chapter 23 **LEGISLATION & REGULATIONS**.....................................**285**
    Regulation 479/2008 Common Organisation of the Market in Wine...286
    Regulation 606/2009 Detailed rules 287
    Regulation 607/2009 More detailed rules...................................287
    Regulation 1991/2004 Declaration of allergens.............................287
    Directive1989/396 Lot marking...288
    Regulation 178/2002 Traceability..288
    Directive 2000/13 Labelling, presentation & advertising..................288
    Weights & Measures (Packaged Goods) Regulations 2006.................289
    Weights & Measures (Specified Quantities) Regulations 2009...........289
    Food Safety Act 1990...290
    Food Safety (General Food Hygiene) Regulations 1995....................290

Glossary...291
Conclusion..297

Acknowledgements

The first seeds for a book explaining the science of wine in layman's language were sown in 1983 by Pamela Vandyke Price, during a tour of the Hungarian wine regions. Pamela cajoled and persuaded me for the whole of the sixteen years that it has taken to finally write the book, and it is to her that I owe the greatest debt. Not only was she the prime mover in the first instance, but she has graciously given the practical assistance that a first-time writer needed.

I am indebted to Kym Milne, a fellow Master of Wine and a renowned modern winemaker, for help with the first edition.

Thanks also to Hugh Johnson, with whom I have worked at the Royal Tokaji Wine Company, for writing the Foreword to this edition.

Grateful thanks are also due to:

Steve Ellis of Sartorius Limited, Keith Pryce of Seitz (UK) Ltd, and Mark Bannister of Carlson Filtrations Ltd for photomicrographs of filter media;

Waverley Vintners Ltd for photomicrographs of yeasts;

John Corbet-Milward of the Wine and Spirit Trade Association for checking the contents of chapter 23;

Taransaud Tonnellerie for information and pictures for chapter 11;

W&J Graham & Co for the picture of their mechanical lagar;

Brian Cane for a most useful and detailed scientific critique;

Elliott Mackey for professional assistance in printing;

and not least to my wife, Alice, for reading the script several times and for detailed checking of the index.

Introduction

This book is aimed at the person with no formal scientific training, yet who is interested in the science behind wine and wants to know the mechanism behind the complex transformations that take place. Scientific terminology has been kept to a minimum and an attempt has been made to use everyday words and phrases. Indeed, there are places where the scientist might raise the eyebrows, places where perhaps science has had to give way to an easy understanding of a complex principle. For this I make no apologies, as this book is not intended as a learned treatise on winemaking and I have put communication of ideas before correct scientific protocol.

Those who have had an education in the arts frequently find that anything scientific is difficult to assimilate. This is a particularly unfortunate in the world of wine, as this is one of the areas of interest where art and science come together. The modern winemaker, graduated from Roseworthy or Geisenheim or Bordeaux, to name but a few of the establishments offering advanced courses in scientific winemaking, must surely also be an artist. There must be an innate feeling for creating something of beauty. As a painter creates a beautiful picture, or a composer an interesting piece of music, so the great winemaker aims to create something which is much more than a mere beverage. Art and science become inextricably linked, which is but one of the reasons that wine is so fascinating.

The great joy of wine appreciation is the infinite variety shown by wines around the world. This is due primarily to the fact that there is no fixed route to the production of good wine. At every stage of winemaking there is a choice and at many of these points totally opposed principles are available, resulting in infinite permutations. It is the correct choice at each stage that separates the brilliant winemaker from the merely good.

In this third edition the information has been brought up-to-date so that this book remains the mainstay for those who are studying for the Diploma Examination of the Wine and Spirit Education Trust or the examination for membership of the Institute of Masters of Wine. The text has also been expanded to include more information on the making of the major styles of the wines of the world, which should be of interest to those who are not engaged in study, but who have an enquiring mind and therefore want to know the mechanisms behind their production. But, above all, I would hope that all lovers of wine will find something of interest, something that will enhance their enjoyment of what is to me the world's best and most healthy beverage.

Foreword

By Hugh Johnson

Call it technology. Call it science: these days we live with it. As a scientific illiterate I was not exactly the most willing participant (and never took an exam). But that is not a viable position any more: you would simply miss too much of the action. We all need a grounding in wine technology to understand what's going on, and those in wine professionally don't get to first base without it.

 I'm not sure whether to call this book a primer, a memory-jogger or a lifesaver. Which it is depends on the reader. For WSET students it is essentially the first, then the second. For people like me it is the third – than rather belatedly the first. What we all need is a crisp exposition of how wine is made and why, easy to refer to when a funny smell appears but going beyond Stinks (do they still call Chemistry that?) to cover the physics, natural history, legislation and finally the appreciation of wine.

David's first edition has been my stand-by for years. I have my Peynaud, my Amerine & Joslyn, my Michael Schuster for going deeper where necessary, but it is always good to have Bird in the hand. This third edition adds a valuable insight into the production of the principal styles of the wines of the world, making it equally interesting for those who are simply lovers of wine and for those who are serious students of the Master of Wine examination. The detailed explanation of the mysteries of Hazard Analysis make this book particularly useful for wineries that are faced with the problems of modern food safety legislation. Essentially, though, it updates the second and makes it available once more to ease the pangs of students young and old.

xx

Chapter 1

THE GIFT OF NATURE

Good wine ruins the purse; bad wine ruins the stomach.

Spanish Proverb

The origins of wine

As even the most moderate of drinkers knows, wine varies enormously in quality, from the positively vile to heavenly nectar. Yet they are all the product of the fermentation of grape juice. The inquisitive wine drinker cannot help but question this diversity of quality and style, and wonder why there should be this wide range, when the raw material is simply the grape.

There can be little doubt that the first wine ever tasted by man was the result of an accident of nature. The vine is an ancient plant and has been known for millennia as the bearer of nutritious fruit. It is highly probable that grapes had long been gathered for consumption as fruit, or for the production of a delicious juice. All that was necessary for the discovery of wine was for a container of juice to have been left standing longer than usual, when the natural yeasts in the atmosphere or on the skins of the grapes would start an alcoholic fermentation and convert the sugars to alcohol. Initially, the juice would have been regarded as spoiled because, in its early stages, fermentation produces an odour of rot and degradation. Indeed, it is the first stage of the degradation process that reduces organic matter to its basic constituents. Only on further keeping, and after a cautious tasting, would it have been discovered that a total transformation had taken place and something had been produced with strange and wonderful properties.

This ancient wine was born without the aid of science and would have been a very hit and miss affair – and ancient it is, going back at least 5000 years, as witness the paintings in the tombs in Egypt. And in the Greek and Roman empires Hippocrates and Pliny wrote about the benefits of drinking wine. It was not until the work of Louis Pasteur (1822 – 1895) at the University of Lille in the middle of the nineteenth century, that it was discovered that fermentation was due to the presence of microorganisms. But it was not until the last three decades of the twentieth century that scientific principles have been rigorously applied to winemaking.

The traditional way of making wine involved little science: grapes were crushed to release the juice, which was allowed to ferment with the naturally occurring micro-organisms until the juice had been converted into wine, with no temperature control and no analysis. The results were totally unpredictable, sometimes wonderful, sometimes disgusting.

The natural cycle

Winemaking is undoubtedly an art and the winemaker an artist, but if an understanding of the basics of the science that lies behind the transformation of grape juice into wine can be grasped, then the full potential of the grape can be realised. The pinnacle of quality can be achieved by the application of science through quality control, used in its holistic sense: controlling quality in the vineyard itself, of the vines, of the grapes, of the expressed juice, of the fermentation process and of the finished wine.

The process begins with photosynthesis, that miraculous process whereby green plants are able to synthesise sugar from carbon dioxide (CO_2) and water (H_2O), under the influence of sunlight and with the aid of chlorophyll, the green matter of plants. The sugar thus generated in the leaves is actually sucrose, which is then transported by the sap to the grapes that are the storehouse of the energy required by the pips to nurture the next generation. When it reaches the grapes, the sucrose is immediately hydrolysed by the acids to glucose and fructose.

As the grapes hang on the vine, and gradually come to full maturity with the aid of sunshine and warmth, the potential quality within the grape reaches a peak. The threat to this quality comes in the instant that man interferes and gathers the grapes. It is critically important that the quality inherent in the grape is maintained between gathering and processing because chemical and biochemical changes occur from the moment the bunch is separated from the vine.

But, more than this, maintaining the quality of the wine after it has been produced is particularly difficult because wine is a meta-stable substance; it is halfway down the slope of decomposition from grape juice to carbon dioxide and water. It is the product of the microbiological attack on grape juice, the same process that reduces all living matter to water, nitrogen, carbon dioxide and a few mineral salts. In reality, wine is a small part of one of nature's cyclic processes, the carbon cycle.

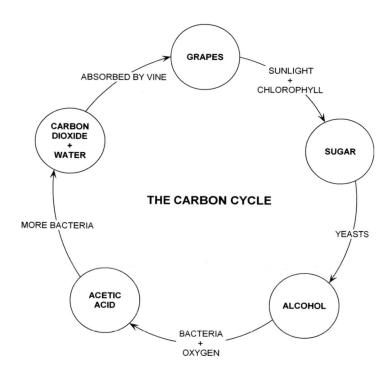

In this process green plants absorb carbon dioxide from the atmosphere and convert it to sugars, which are then used to create alcohol. This in turn gives rise to vinegar, which itself decomposes to release carbon dioxide into the atmosphere, which starts the cycle all over again.

The complete cycle in detail is as follows:

1. The vine absorbs carbon dioxide (CO_2) from the atmosphere through the stomata in the leaves and water (H_2O) through its roots. With the aid of sunlight and the green chlorophyll in the leaves, it converts these raw materials to sugars, which are stored in the grapes.

2. After the grapes have been picked and crushed, the yeasts convert the sugars to alcohol, producing wine, with some of the carbon atoms from the sugar returning to the atmosphere as carbon dioxide.

3. The next stage is the process known as oxidation, when the wine is attacked by the oxygen of the air, with the aid of bacteria, which converts the alcohol to acetic acid, producing vinegar.

4. By the action of more bacteria the vinegar is decomposed, yielding, ultimately, water and carbon dioxide.

5. The water flows into the soil and the carbon dioxide goes into the atmosphere where they are ready for further re-cycling.

Enzymes in nature

All of this activity is dependent upon the presence of enzymes that are the key to all life processes: if they are poisoned, life ceases. The reason that cyanide is such a deadly poison is that it inactivates the enzymes in our bodies, thus stopping all our essential bodily processes and death is almost instantaneous. An enzyme acts as a catalyst, a substance that promotes a particular chemical reaction but does not actually take part in it. They are large and complex molecules that are one step down from living organisms; they are not actually alive and cannot reproduce themselves, but they can easily be poisoned (see p.178 for more).

Enzymes in wine need controlling, for there are both good enzymes and bad enzymes. The good ones are responsible for the process of fermentation, while the rogues (known as the oxidases) assist in rapid oxidation. Fortunately, it is not necessary to resort to cyanide to control them, as there are many less noxious chemicals that will inactivate them, sulphur dioxide (SO_2) being the one that is widely used in winemaking.

The application of science has given the winemaker choice; at every stage of the winemaking process there are options, many of which have only become available because of scientific research. The potential for making good wine is better than ever before. Equally, never before has such a range of high quality wines been available at such low prices.

Wine and health

The benefits of drinking wine first came to light in 1991 in an edition of the *60 Minutes* programme on CBS that discussed the results of a study that came to be known by the colloquial title of The French Paradox. The purpose of this study was to demonstrate that dairy fat consumption and coronary heart disease are highly correlated. However, the paradox occurred amongst the population of a few French cities which showed that, despite a high proportion of fatty foods in their diet, the incidence of coronary heart disease was low. Further investigation indicated that the consumption of alcohol played a large part in this situation.

In 2008 it was reported in The Times that the death rate for coronaries among men aged 35 to 64 in Toulouse was 78 per 100,000, compared with Belfast where it was 348 and 380 in Glasgow. The average drinker in each city consumed about the same amount of alcohol, but in Toulouse it was almost exclusively red wine.

In the first edition of this book, published in the year 2000, the opening sentence read: "What a joy it is to be living in an era when we are told that wine is good for us!". How opinions have changed in the past decade! In earlier years we had the constant warnings of the dangers of alcohol, how it is addictive, how it can damage the liver, how families are ruined by excessive drinking. These dangers are still very real and moderation remains the principle by which we should all abide. The problem now is that we are receiving mixed messages, especially from the multitude of research groups that are studying the effects of alcoholic beverages on the human body. This is a popular field of research because the participants know that the results will be leapt upon with glee by the media. The unfortunate result of this deluge of research data is that conflicting results are being published. One group finds that drinking wine causes cancer, whereas another group maintains that the polyphenols in the wine protect against cancer. Another declares that wine increases blood pressure and is bad for the heart, and a fourth group states that wine contains substances that protect the heart. The best advice is to ignore most of these research results and just enjoy moderate drinking. After all, wine has been consumed for centuries and was always regarded as a healthy, regular beverage, provided it is consumed with moderation.

Spirits and binge drinking are two of the greatest dangers. The high alcoholic strength of spirits causes the level of alcohol in the blood to rise higher than the equivalent amount of alcohol taken in a more dilute beverage, such as wine or beer. Regular moderate drinking results in a rise in the level of alcohol dehydrogenase (the enzyme that breaks down alcohol) in the liver, thus helping the body to metabolise the alcohol efficiently and quickly. With binge drinking, this does not occur. The body struggles to cope with the sudden deluge of alcohol, resulting in severe intoxication and damage to the organs.

The attributes of the specific beneficial components of wine listed below have been shown by many studies to be factual and are not the result of spurious and limited research.

- *Alcohol*

It has long been known that alcohol increases the high-density cholesterol, the so-called "good cholesterol" that lowers the risk of heart disease, at the same time lowering the dangerous form, the low-density cholesterol.

Alcohol also plays a positive role by acting as an anticoagulant, guarding against thrombosis by preventing the aggregation of platelets in the blood. It also has a relaxing effect on the system and can be of great social value, provided it is not abused.

- *Flavonoids*

Recent studies have shown the flavonoids (polyphenols, see p.25) to be an extremely important group of compounds, particularly in relation to the human body and health, because they are powerful antioxidants, as are several of the vitamins. The ageing of the body is largely due to oxidation, so the presence of antioxidants in the diet is essential. Some members of the flavonoid group also share with alcohol the valuable property of preventing the clumping of platelets in the blood, which reduces the chance of coronary heart disease. Thus it can be seen that these flavonoids confer on wine that precious quality of 'being good for you', actually slowing the process of ageing. This applies particularly to red wines, as they contain higher levels of polyphenols than white or rosé wines. Hence, the recommendation that two glasses of red wine per day are beneficial!

•   *Resveratrol*

Resveratrol is another health promoting substance that has recently been discovered in wine. It is a member of a group of compounds known as phytoalexins that are produced in plants during times of stress, such as bad weather or insect or animal attack, and help to protect them from fungal disease. It can be found in many plants, but red wine has been found to be a particularly rich source. It is known to be both antioxidant and anti-mutagenic and inhibits all three phases of the cancer process: initiation, promotion and progression. Further, it appears to have a certain amount of antibiotic action, and can control the growth of unpleasant bugs, such as *Chlamydia pneumoniae* and *Helicobacter pylori*.

•   *Potassium*

Another positive aspect of wine in relation to health is the high natural level of potassium salts that it contains, one of the highest of all foodstuffs. Potassium is valuable in counteracting excess sodium in the body. Although sodium is an element that is essential to biological functions, an excess causes an increase in blood pressure, or hypertension. Modern diets, especially snack foods and many fast foods, contain too much sodium for good health. Potassium, although chemically closely related to sodium, does not have this bad effect on blood pressure, but does have the useful property of replacing sodium in the body, the sodium being excreted in the urine.

•   *Histamine*

Histamine is one of a family of what is known as biogenic amines. Amines are one of the essential building blocks of living organisms, and play an important role in life as we know it on Earth. Biogenic amines have a bad reputation for causing problems such as low blood pressure, facial flushing, nasal congestion and other forms of distress. Wine has sometimes been blamed for the appearance of these symptoms, but it is highly unlikely that this has been due to the histamine content of the wine. Foods such as mature cheese, fish, meat and yeast extract contain at least ten times the level of histamine found in wine.

The modern paradox

Governments in many countries struggle to find a way of reducing the harm that is done to the human body and to communities in general by the excessive consumption of alcohol. Most of the measures suggested are fiscal, such as increasing the tax or applying a minimum price control per unit of alcohol. It is most unfortunate that these measures punish the majority of regular, moderate drinkers, while having little effect on the cause of the problem. New regulations are not necessary, as there are plenty of existing laws that prohibit the sale of alcohol to juniors and to those who are intoxicated. The problem is that these laws are difficult to enforce. The only real answer is to change our attitude towards drunkenness, and this will not be easy. We need to convince young people that it is not clever or cool to be drunk: it is pathetic and stupid.

The beneficial properties of wine are manifold: it can lower the risk of coronary heart disease, thrombosis, Alzheimer's, cancer and leukaemia, and it protects against the modern scourge of eating too much salt.

So, let us revel in regular, moderate drinking (up to half a bottle of wine per day for a man, and slightly less for a woman) and gain the benefits that wine can bestow. It is, after all, one of the oldest beverages known to mankind.

Chapter 2

IN THE VINEYARD

Noah, a tiller of the soil, was the first to plant the vine.

Genesis 9 [20]

The vine

Wine is made from grapes, and only from grapes. This may sound a somewhat dogmatic statement, especially to those who are used to making 'wine' from apples, raspberries, currants, rhubarb, nettles, or indeed any edible fruit, vegetable or leaf. Such products must be named after the substance from which they are made, such as 'Apple Wine'.

Grapes are the fruit of the vine, which is a member of a large family of climbing plants, only a few of which are suitable for making wine. The study of the vine is known as ampelography, coming from the Greek *ampelos*, which means, not surprisingly, the vine.

The **family** to which the vine belongs is known as *Vitaceae*, plants that show a tendency towards trailing and twining. This is a large family of eleven genera, but by far the most important is the **genus** *Vitis*, the grape vines. Another well-known genus is *Parthenocissus*, the genus to which the Virginia Creeper and Boston Ivy belong.

The genus Vitis contains some 60 different species, few of which are suitable for the production of grapes for making wine. The most important **species** by far is *V. vinifera*, the European grape vine. Other species of use in winemaking are: *V. labrusca,* the North American Concord grape; *V. riparia,* the Riverbank grape; *V. aestivalis,* the Summer grape and *V. rotundifolia,* the Fox grape. *V. vinifera* is the only species that gives rise to wine of a flavour that is acceptable around the world, but some of the other species are important in providing root-stocks on which to graft the European vine in the fight against phylloxera (see p.11).

Within the single species of *V. vinifera* there are numerous **varieties**, probably around one thousand, many with well-known names such as Cabernet Sauvignon and Chardonnay, probably the two most widespread of the international varieties.

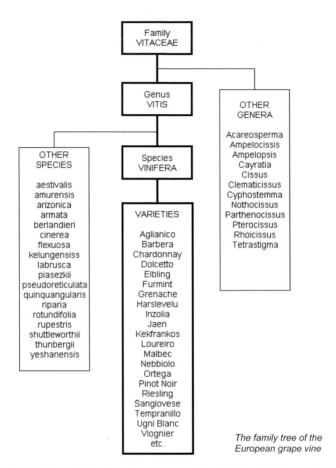

The family tree of the
European grape vine

The selection of the right vine does not end with the choice of a variety, for it is now possible to select **clones** of many of the better known varieties. This use of the word clone is somewhat misleading, its true meaning being an identical copy. In the case of vines, it actually means the selection of a particular vine that has characteristics that set it apart from the majority. When this vine has been selected, it is propagated by taking cuttings. It is possible to select clones specially bred for high quality, disease resistance, higher yield, deeper colour or smaller berries. Therefore clones are different from the general population, but are identical to each other within the clonal group.

Genetic modification is an obvious step to be taken at some time in the future. The vine is a delicate plant, subject to attack by many agents such as moulds, mildews and viruses. A vine that has a built-in resistance towards such troubles would be widely received – but only after a sane and balanced attitude towards GM plants has been established.

Phylloxera & grafting

It must be to the permanent shame of England that the pestilence of *Phylloxera vastatrix* was introduced to Europe via Kew Gardens, albeit unwittingly. Vine plants brought over from America in the nineteenth century for experimental purposes were carrying an infestation of Phylloxera that rapidly spread throughout Europe, destroying many of the long-established vineyards by attacking the roots. It must have been devastating to watch vines dying mysteriously year by year, not knowing the cause of the problem. Towards the end of the nineteenth century it had decimated 70% of the vineyards of Europe and it really looked as if wine production was coming to an end. It was eventually discovered that the cause of the problem was a wicked little aphid with a complicated life history. The adult form is a winged insect that can fly from vine to vine. This insect lays several hundred eggs, which hatch into larvae, some of which go down into the soil and attach themselves to the roots where they form nodes. This soon results in the

Vine grafts awaiting planting.
The bottom part is the
American rootstock
and the European bud at the top
has been coated in green wax
to prevent it drying out.

complete destruction of the root system. After many desperate measures were taken to eradicate the pest, such as flooding the vineyards, or treating with hydrogen sulphide (a particularly noxious and dangerous chemical), it was beaten by sheer logic. Scientific detective work traced the origin of the pest to America, where it was living happily with the American native vines. This being the case, it seemed a logical step to graft the European vine on to American rootstock, with complete success.

Grafting has become the norm throughout the world, with the exception of Chile and Cyprus where the pest has never been able to reach the vineyards due to the barriers of oceans and mountains. Grafting has actually yielded a positive benefit, in that it is possible to select a suitable rootstock for matching the vine to the soil. In rich soils a weak rootstock can be used; in chalky soils an alkaline-resistant rootstock is chosen. The choice of rootstock is now regarded as important as the variety of the vine itself.

Climate

The balance between sugars and acids in the ripe grape plays an important part in the quality of the finished wine. This balance is determined by the degree of sunshine that the grape receives during the growing season, which is defined at a basic level by the climate: with too much sun the acids fall too low and the sugars climb too high, resulting in a flabby alcoholic wine. Conversely, too little sun gives a thin, sharp wine, and possibly with insufficient alcohol even to be called wine, because the EU regulations for table wine set a minimum limit of 8.5% for the alcohol produced from the natural sugars.

Given that it is not within our power to control the climate, we can influence the ripening of the grapes by paying attention to microclimate and by the careful choice of elevation and aspect (the direction in which the vineyard faces, relative to the sun).

Microclimate plays a significant part in the successful production of wine in many parts of the world, either by reducing the effect of the sun by going up to higher altitudes, or, in areas that suffer from a deficiency of sunshine, by using slopes of the right aspect to make better use of the sun's rays, especially during the morning or the late afternoon.

Training & pruning

The various shapes into which a vine can be trained give us another way of controlling the effect of heat on the ripening of the grapes. The arrangement of the leaves forms what is known as the canopy of the vine. The manipulation of this canopy, or canopy management, is particularly important in marginal climates, where the different styles of training can be used to either maximise or minimise the effect of the sun's rays. In vineyards such as Châteauneuf-du-Pape the effect of the sun is deliberately emphasised by training the vines in the bush style, with short stems, so that the producing parts of the vine are close to the ground. This traps the maximum warmth from the sun's rays, both directly and from the reflected heat from the *galets* or 'pudding stones'. This effect continues into the night, when the stones act as gigantic storage heaters, enveloping the vines with warm air.

In hot climates where a lighter, crisper wine might be desired, the vines are trained high, with their upper branches formed into a pergola. In this formation the grapes hang down underneath the foliage and remain in the comparative cool of the shade.

Grapes in the Marche region of Italy trained in the tendone style to reduce the effect of the sun

Conversely, in a cool marginal climate such as England, much use is made of specialised training methods such as the Geneva Double Curtain, where the ratio of foliage to grapes is carefully controlled to give the maximum exposure to sunlight and, thus, the maximum production of sugar.

The principle of maintaining the concentration of the flavour elements in the juice applies equally to the annual round of pruning. This needs to be severe, so that the sparse supply of the precious constituents is not spread too widely over a large number of bunches, but shared between the lucky few. Grapes are the basic material from which wine is made, and the potential quality of the ultimate wine is already present in the grape juice. This quality must be preserved throughout the winemaking process. The adage "Fine wine is created in the vineyard" is very true. Or, again, "It is easy to make bad wine from good grapes, but you cannot make good wine from bad grapes".

Soil & water

The basic principle of growing grapes for winemaking is to force the vine to develop a large underground system of roots that deliver an ample supply of minerals and nutrients to the growing leaf system, where the sugars are being generated by photosynthesis. All of these components are eventually transported to the grapes, which act as storehouses. Vigorous pruning of the above ground system ensures that these substances are concentrated into a small number of grapes. In rich soil the vine produces enormous top growth from a small root system and the grapes never reach the intensity of flavour that is required for good wine. In well-drained poor soil the vine is forced to develop a large root system that penetrates deep into the sub-soil in search of moisture and nutrition, and in so doing it picks up an abundance of minerals that find their way into the grapes.

There can be no denying that the vine reacts in a similar way to humans in relation to performance: the best results are obtained when subjected to a certain degree of stress. The idle and cosseted person rarely produces anything of interest and, similarly, the vine in rich soil yields much foliage and fat grapes, which produce dull wine. This was precisely the situation in the middle of the last century, when Aramon grapes were planted in the rich soil of the plains of the Languedoc. The resultant wine was of a very poor quality and was virtually impossible to sell.

In common with many other herbaceous plants, the vine has no tap root, but has a large number of fine feeding roots which can burrow down tens of metres into the subsoil when forced to do so by lack of moisture. The drainage of a vineyard is one of the major influences in the production of quality grapes. It was discovered some years ago that the difference in quality along the gravel slopes of the Médoc was due largely to the proximity of small streams and man-made drainage channels that allow the water to flow out of the gravel, the best vineyards being the most well drained.

Although the importance of good drainage is widely accepted, the effect on quality of the various soil types is not fully understood, probably because the vine will tolerate a wide range. The broad principles that have emerged are that alkaline soils emphasise the characteristics of white grapes, as in Sancerre and Champagne, and gravel soils produce particularly good red wines, as exemplified by the Médoc. But the relationship between soil and vine variety is far from rigid, and fine wines are produced around the world on various combinations of soil and vine.

The concentration of essential elements in the juice is also related to the density of planting. The underlying principle is to make the vine act as a concentrating plant, gathering the minerals and flavour precursors from a large volume of strata and sub-strata and concentrating them into a small number of grapes. Planting density provides a useful way of controlling the division of the available resources in the soil among the right number of plants. In a rich soil, vines are planted closer together, resulting in competition for the nutrients, which makes the roots delve deep into the sub-soil.

A technique that has gained favour recently is the use of herbaceous plants between the rows to control the moisture that is available to the vines. Time was when it was regarded as poor management to allow weeds to grow between the vines. Now it is realised that such plants will compete for moisture and nutrients in a manner which has a positive effect on the quality of the grapes. In a year when there is an excess of rain, the herbaceous plants will remove a proportion of the moisture from the soil, resulting in the grapes being more concentrated. Some vineyard owners select the precise type of plant and will even sow between alternate rows of vines. During the growing season the herbaceous covering is mown and the mowings are left in place to decompose, with the valuable nitrogen being returned to the soil.

Irrigation

Irrigation is a technique that is prohibited in most of the vineyards in Europe that are producing quality wines, the reasoning being that irrigation results in a dilution of the juice in the grapes. This is a pity for several reasons.

In a dry season, a judicious application of water would result in a sensible quantity of well-balanced wine. Without the water, the vines tend to shut down due to hydric stress, and the resultant wine is tannic and lacking in fruit.

Many vineyards in the New World use irrigation, very often as routine. This is an example of the Europeans 'shooting themselves in the foot' by retaining such restrictive legislation.

There is a clever technique known as 'partial root drying', whereby the roots of the vine are irrigated on one side only, the other side remaining dry. The dry roots send a chemical message to the plant to shut down because of hydric stress, so it switches its energy into fruit growth rather than the production of vegetation. But the wet side of the roots keeps supplying water, which goes straight into the development of the grapes.

The control of moisture in the vineyard soil is undoubtedly a useful way of manipulating the growth of the vine and the production of quality grapes. The overuse of irrigation results in dilute juice and poor wine, a situation that benefits nobody.

Green harvest

The aim throughout the growing season is to achieve grapes that are healthy and concentrated, rather than fat and luscious like table grapes. This involves watching the vines for correct growth of the canopy, with adequate leaf development to ensure good sugar production, but avoiding too much shading of the grapes that would result in a green and stalky flavour. So careful summer pruning is usually necessary.

The number of bunches per vine is important, bearing in mind that the vine is acting as a concentrating plant, pushing all of its energy and flavour into a small number of grapes. It is not easy to forecast the yield at the time of pruning, so it is often necessary to perform what is known as a 'green harvest'. This apparently wasteful process involves cutting off a proportion of the bunches while they are still green. These grapes are just left on the ground to rot, making the vineyard look terrible. However, it is a useful process, resulting in more concentration in the remaining bunches.

Bunches of grapes are cut out and left on the ground where they will rot

This has to be done at the correct time for it to be effective. If too early, the vine senses that it has lost a lot of fruit and it tends to start shutting down, which defeats the object of the exercise. If too late, then much of the energy that has been put into the grapes is lost. Usually, the correct moment is around the time of *véraison*, or the changing of the colour.

Terroir

Much has been said and written about terroir, a French word that has the literal meaning 'relating to the earth'. This, however, is not the way in which the word is used with reference to wine production, where it means far more than earth. It is more related to the way the geography of the vineyard affects the quality of the wine.

Terroir includes climate, soil type, topography and any other influence, even if human. It really means the totality of all the outside influences that can alter the quality of the grapes and hence the wine.

It has been decried by many people who maintain that it is a fancy word coined by the owners of the best sites merely to protect their value. This may be the case, but there is strong evidence to suggest that there is a great deal of truth in the concept.

Vineyard systems

The vine, although vigorous in growth, is actually a delicate and sensitive plant that is prone to all manner of troubles and diseases. In many parts of the world this is tackled in a somewhat cavalier fashion by regular spraying with various natural and manufactured substances designed to kill the attacking organism. This approach is crude and damaging to the environment and is gradually giving way to systems that are more sympathetic and take the approach of working with the vine rather than against it.

• Viticulture Raisonée (La Lutte Raisonée)

'The reasoned fight' is an approach where the vine and its environment are respected and are treated only when necessary to maintain the fight against pests and diseases. In France this is supported by the Terra Vitis organisation.

Its basis is observation of the vineyard and monitoring of the state of the vines. A certain level of pests and diseases is accepted, and only when this level is exceeded is any treatment given.

Records of all treatments must be maintained, and traceability throughout the process must be possible.

This seems an eminently sensible principle to follow as it fosters a much closer relationship between vineyard and viticulturist, and it avoids the unnecessary application of chemicals that occurs when blindly following a calendar routine.

•   *Organic viticulture*

Although one often hears the phrase 'organic wine', it has no real meaning. The correct definition of wines made by this method is 'Wines made from organically grown grapes'.

Entirely natural methods are used for controlling insects, fungus and weeds. Forward-looking, natural farming techniques are at the heart of all of the system. No herbicides, insecticides, pesticides, or chemical fertilizers are used, and only approved, naturally occurring substances are applied. Sulphur dioxide is allowed as an antioxidant and antiseptic, but at a lower level than in other wines.

Organic vineyards have to be separated by areas of natural vegetation or forest to eliminate the risk of contamination from the chemicals used in other vineyards.

With certified organic products, there is an assurance that:

1. no harmful synthetic chemicals have been applied to the land for at least 3 years;
2. only non-toxic, environmentally friendly methods and materials have been used to grow the crop;
3. non-toxic equipment sanitisation and pest-control methods have been used;
4. there has been no exposure to prohibited materials during bottling.

The entire process, from vineyard management and grape processing to the final bottling, comes under the scrutiny of an accredited body in the country of origin, with an annual inspection.

•   *Biodynamic viticulture*

This system is the ultimate state of farming with nature, rather than relying on man-made interference. Its principles were first elucidated in 1924 by Rudolph Steiner, an Austrian philosopher and scientist. The key to its operation is to see the whole farm as a single living organism, where wrongful interference by man at any stage can result in sickness. Artificial fertilisers or pesticides are total anathema. Control is by the use of all things natural, in conjunction with the correct timings according to cosmic rhythms.

It is very easy to pour scorn on this method, as it makes use of somewhat unusual substances:

- cow manure fermented in cow horn;
- flower heads of yarrow fermented in stag's bladder;
- stinging nettle tea;
- juice from valerian flowers;
- infusion of horsetail plant.

The First International Biodynamic Wine Forum was held in Australia in 2004 at which the keynote speaker was Nicholas Joly of Coulée de Serrant. The brochure gave the following as an explanation of Biodynamics:

> The Biodynamic method involves the use of specially developed preparations that assist in connecting the whole farm unit with the dynamic rhythms of the earth and atmosphere.
>
> Instead of just acting on the physical, Biodynamics goes one step further in both working with the living soil and the invisible energies of nature.
>
> Because of this connection with this world of energies, Biodynamics helps to dramatically increase the possibility of individuality, an individuality the French call *terroir*.

However difficult it might be to accept some of the fundamental principles of biodynamic viticulture, the undeniable fact is that excellent wines are produced using this system. Famous biodynamic producers include Domaine de la Romanée Conti and Domaine Leflaive in Burgundy, Zind Humbrecht in Alsace, Joly in the Loire Valley, Chapoutier in the Rhône Valley, and there are many others in countries around the world.

Chapter 3

INSIDE THE GRAPE

My friend had a vineyard on a fertile hillside . . .
He expected it to yield grapes, but sour grapes were all that it gave.

Isaiah 5.1-2

Grapes supply all of the components of wine that give it its ultimate quality and style. The winemaker can influence the final balance, but he cannot add quality if it is not there in the first place. The balance of the sugars, acids and polyphenols in the grape juice is of prime importance to the style of the finished wine. Too much sunshine and heat will yield a wine that is highly intoxicating and short on acidity – hot and flabby. The converse produces an unpleasant, thin and acidic concoction that is fit only for distillation (and, incidentally, might make a wonderful base wine for the production of a top-flight brandy). 'Balance' is a word used frequently throughout the production and tasting of wine, because it is the all-important balance of the major components of juice that yields an attractive wine.

Sugars

In common with most fruits, unripe grapes contain low levels of sugars and a high concentration of acids: sunshine and warmth are required for the production of adequate sugars by the process of photosynthesis. This is the amazing biochemical process by which green plants are able to synthesise sugars from carbon dioxide (CO_2) and water (H_2O) with the aid of chlorophyll and energy from sunlight and warmth. The sugar produced by this process is sucrose, the same sugar that we extract from sugar cane or sugar beet.

$$12CO_2 + 11H_2O \rightarrow C_{12}H_{22}O_{11} + 12O_2$$
sucrose

It can be seen from the above equation that this process not only produces oxygen, but also absorbs carbon dioxide, one of the planet's greenhouse gases. This is the reason that the destruction of forests around the world is said to be adding to the danger of accelerating global warming.

Throughout the growing season, the sucrose manufactured in the leaves of the vine is transported through the plant to the grapes, which act as storehouses. But a major change occurs within the grapes. Due to the presence of acids in the grape, the sucrose is immediately broken down (hydrolysed) into glucose and fructose.

$$C_{12}H_{22}O_{11} + H_2O \rightarrow C_6H_{12}O_6 + C_6H_{12}O_6$$
$$\text{sucrose} \qquad\qquad \text{glucose} \quad \text{fructose}$$

Glucose and fructose are closely related in that their molecules contain the same number of atoms, but in a different arrangement. In chemical terms these are known as isomers, and they have different properties, especially relating to sweetness. Fructose is the sweetest, followed by sucrose, with glucose being the least sweet. Wine yeasts actually prefer glucose, which they will consume before fructose, leaving the fructose as the principal sweetener in sweet wines.

The progress of the build-up of sugars in the grape is very weather dependant: the higher the temperature and the greater the sunshine, the greater the level of sugar produced.

It is easy to determine the progress of sugar production by measuring the sugar content of the juice with a pocket refractometer. This is a simple instrument that gives a direct reading in terms of sugar content, but actually works by measuring the refractive index of the juice. This measurement gives a good indication of the sugars because they are present at a concentration at least ten times that of the other constituents. A few drops of juice are placed on the prism, the top closed and the scale read by peering through the eyepiece.

Refractometer scale giving a reading of 365 g/l of sugar in aszú berries

Refractometer with prisms open showing where drops of liquid are placed

The perennial problem is the lottery of the weather. If the sugar level is low, further ripening on the vine might result in an increase in the sugar content; but if it rains, the sugar concentration might fall as a result of dilution. The correct decision at this moment lies with the expertise of the winemaker.

Acids

The second principal component of most fruits is the acids. Grapes contain natural acids that impart freshness and keeping qualities to the finished wine, and are an essential component of the taste of all wines. Contrary to the production of sugars, the acids are produced mainly in the grape itself.

The two main acids found in grapes are malic and tartaric, which together constitute over 90% of the acidity of the grape.

Malic acid $\quad$ $HOOC.CHOH.CH_2.COOH$

Tartaric acid $\quad$ $HOOC.CHOH.CHOH.COOH$

The very young, green grape contains mostly malic acid, the principal acid of apples (L. *malus* = apple), and has a very sharp taste. It plays an active role in the life processes of the grape and even in the subsequent wine. This is the reason winemakers are sometimes wary of the presence of malic acid in the finished wine, and prefer to use tartaric acid for any adjustments.

Firstly, the malic acid is consumed by the grape as an energy source. Then, at a later stage in the maturation pathway, another mysterious transformation occurs. The malic acid is able to undergo a conversion to glucose, which to a layman may seem surprising. This is known as gluconeogenesis – the new production of sugar (*gluco* = sugar, *neo* = new, *genesis* = creation). This is an important process in the human body, as glucose can be synthesised from other substances when the body is under conditions of starvation or excessive exercise.

The net result of this activity is that the proportion of malic acid in the grape goes down as maturity approaches.

Tartaric acid is quite different and is formed as a by-product of the synthesis of sugar. It is the main acid found in most finished wines and is unique to grapes. It gets its name from the salts which form the major proportion of the deposits found in containers after the storage of wine (L. *tartarum* = deposit). Tartaric acid is biochemically a fairly stable acid, and thus the quantity in the grape rises in proportion to the creation of sugars.

Therefore, as the grape ripens, the proportion of malic acid goes down, while that of tartaric acid rises.

However, the total acidity of the juice in the grape goes down, and this fall in acidity is greater with increasing temperatures. The prime reason for this is simply the dilution of the acids by the large quantity of sugars that are being transported to the grape. Thus, the acidity due to the tartaric acid falls because of dilution with sugars, but the level of malic acid falls even further as it is destroyed biochemically.

At a later stage of wine production, after the alcoholic fermentation, the malic acid can undergo another biochemical conversion, this time with the aid of a bacterium and known as the malo-lactic fermentation. In this case the very acidic malic acid is converted into lactic acid, a softer and more gentle acid that occurs in sour milk (L. *lac* = milk) (see p.84).

Acids are easily measured by a simple titration (see p.235)

Mineral salts

Grape juice is rich in many different minerals, picked up by the roots of the vine as they delve deep into the subsoil in search of moisture. The most abundant of these by a factor of at least ten are the salts of potassium (K), which are associated with the production and translocation of the sugars. As the sugars are the most abundant constituent of grape juice, it is not surprising that potassium is present in such quantities. As would be expected, its concentration rises with the accumulation of the sugars in the grape.

It is, however, something of a double-edged sword. On the one hand, it is a valuable element, conferring on wine health-giving properties. On the other it is the cause of many problems relating to tartrate crystals, for it is potassium bi-tartrate that precipitates in bottles, resulting in consumer complaints.

The next most abundant element is calcium (Ca), which also plays a part in tartrate crystal formation. Other elements include magnesium (Mg), with iron (Fe) and copper (Cu) at even lower levels, and traces of many others.

The total mineral salt content of the juice plays an important role relating to the acids because these salts affect the acid taste of the finished wine. Although the acids in the juice supply the acidity, the actual taste in the mouth is determined not by the level of the acids, but by the pH of the juice (see pH on p.237). The pH is controlled by an interaction between the acids and the salts. The presence of certain salts can change the degree of acidity of the acids, an effect known as 'buffering'.

Phenolic compounds

This is a complex and wide-ranging group of compounds that can be divided into two main classes, the non-flavonoids and the flavonoids.

Non-flavonoids are simple compounds with comparatively small molecules based on benzoic acid and cinnamic acid, residing in the pulp of the berry. Their influence on the flavour and taste of the finished wine is relatively minor.

Flavonoids (polyphenols) are much more complex with larger molecules and are found in the cells of the skin and stems. One of the most important properties of the polyphenols in general is that they are powerful antioxidants, which act as preservatives both for the wine itself and for those who consume the wine. One of the reasons that red wines in general have a longer life than white wines is that they contain higher levels of polyphenols. Similarly, red wines are said to have greater beneficial effects on the human body than white wines.

Polyphenols can be sub-divided into tannins and anthocyanins.

• *Tannins*

Tannins are a widely distributed group of compounds found in many plant materials. Probably the best-known sources are tea and rhubarb, whose drying effect on the palate is well known. In grapes they are found in the skins, the stems and the pips. The principal source is the skins, where the tannins are held in fairly tough cells in the outer layers of the skin. If skins are treated roughly or pressed too hard, too many of these cells are ruptured, resulting in too great an extraction of the tannins.

Stems are rich in tannins, so leaving them in contact with the juice during the maceration process results in a juice with higher tannin content. On the other hand, if the skins and juice already contain an adequate supply of tannins, de-stemming is carried out on either the total crop or a proportion of it.

The seeds also contain tannins, amounting to between 20 and 55% of the total polyphenols of the berry, hence the importance of physiological maturity (see below).

With increasing warmth and ripeness, the chemical composition of the polyphenols in the skin changes, with the result that the harsh and 'green' tannins in the unripe grape become softer and more approachable. This phenomenon plays a big part in the taste characteristics of red wines from different climates. Wines from cool regions such as the Loire valley and Bordeaux have a typically hard tannic taste, whereas those from hot countries such as Australia and California have softer, riper tannins.

• *Anthocyanins*

The anthocyanins are responsible for the colour of red wine, and are found in the softer cells towards the inner layers of the skin of the grape. Thus, fortuitously, they are extracted more readily than the tannins during the maceration process. Most black grapes have colourless juice, so this maceration is essential for the production of well-coloured red wine. There are a few that have red coloured juice, known as teinturier varieties, such as Alicante Bouschet, but these are the exception.

The depth of colour of red and rosé wines depends entirely on the way in which these substances are extracted from the skins during the fermentation process. In traditional fermentation, the alcohol produced by the yeast extracts the anthocyanins from the skins, thick skins yielding more colouring matter than thin skins. This extraction can also be achieved by the effect of heat, which breaks down the cell walls (see p.96).

The anthocyanins are one of the groups of compounds responsible for the colours of fruits and flowers and have been named after the principal group of plants to which they give their colour: malvidin (mallow - purple), delphinidin (delphiniums - blue), peonidin (peonies - pink), cyanidin (cornflowers - blue). During maturation of the wine the ratio of these various anthocyanins changes, the proportion of the blue-coloured compounds diminishing, hence the gradation from purple-red to orange with age.

The chemistry of the whole polyphenol group of compounds is complex indeed. In young red wines, the colour is mainly due to the anthocyanins themselves, which are not particularly stable. As the wine ages, anthocyanins link with tannins to form pigmented polymers, which are more stable. And then there are further reactions between anthocyanins and other phenolic compounds, and between anthocyanins and aldehydes. Many of these more complex compounds are more stable than the original anthocyanins. Much of this chemistry is not fully understood, but it is now realised that the managements of polyphenols is one of the major keys to good winemaking.

Flavour components

Sugars, acids and polyphenols are important groups of compounds in the grape and represent three of the four basic tastes: sweetness, sourness and bitterness. If this were the entire collection of substances to be found in a grape, all wines would be very similar and very boring. There is another group that confers the individual flavours of each variety and each style: the flavour compounds.

This is an extremely complex and wide ranging group of substances, present in minute quantities which can only be measured in parts per million, parts per billion, and some even in parts per trillion. Together they form what is known as the 'primary aroma' of the wine, giving it its varietal character.

These compounds are contained in the cells that form the inner surface of the skin, hence the importance of the skin and the role it plays during the processing of the grape. The close association of these aroma compounds with the skin of the grape has encouraged many winemakers to experiment with various forms of skin contact, especially with white grapes, in order to extract maximum flavour and character. (See p.104)

It is the concentration of all these components that creates fine wines. The complexity of these components makes it almost impossible to define the quality of a wine in terms of chemical analysis. The simple analysis of Château Latour of a good vintage, for example, would be identical to that of a basic vin de table. The difference lies in neither the alcohol, nor the total acidity, nor any of the other basic components, but in the myriad minor constituents that make up the flavour of the wine.

These substances include complex esters and higher alcohols, aldehydes, terpenols and hydrocarbides. In some cases, the aromatic substance is present *per se*, others are precursors of the ultimate compounds, and some are unstable and are transformed into other odorous compounds.

If it were important to reduce the assessment of quality to chemical terms, it would be necessary to expand the analysis to include several hundred compounds, a very expensive and time-consuming task. Even the new 'mechanical noses' that have been invented find it impossible to differentiate the vast range of qualities that are available. A trained taster is able to do the job much more quickly and more cheaply, and the wine gives pleasure in the process – which, after all, was the original intention of the winemaker!

Proteins & colloids

Proteins and amino acids, the building blocks of proteins, provide sources of nitrogen for the growing yeasts during fermentation. Glycerol (glycerine), a slightly sweet tasting syrupy substance when in concentrated form, is present partly in the original juice, but is also formed during fermentation and confers a smoothness and body to the finished wine.

All of these substances play a large part in the formation of complex flavours by inter-reacting and producing a vast number of new compounds during the maturation process of the wine.

Some of the largest molecules in grape juice are the proteins and other colloids that are essential to the health of the grape. Proteins are built up from amino acids, which are the basic building blocks of all organic life. Because of the size of their molecules, which are larger than simple matter such as glucose by a factor of around a thousand times, they are classified as colloids. They are nutritious, both to yeasts and to the ultimate consumer, but they have one drawback in that they cause the wine to become hazy or even cloudy after a certain period of time, a period which can vary greatly from a few weeks to many months.

Colloids have very complicated molecular structures, with molecules that appear under the electron microscope as tangled chains of carbon, hydrogen, oxygen and nitrogen atoms. Some of these colloids are totally stable and never change their properties. Others, known as the unstable colloids, slowly change their nature as their molecules re-arrange themselves (known as denaturing) and cause deposits in the finished wine. In old-style winemaking this occurs naturally during the long period of maturation in the cask. But the speed with which modern wine is rushed through the process necessitates the removal of unstable proteins by the treatment known as 'fining' which is explained on p.143.

Véraison & maturity

There is more to the maturation of the grape than merely the change in the acid/sugar balance. It also involves many other processes including changes to the polyphenols in the skins.

One of the major points in the life of a grape is known by the French word *véraison,* for which there does not appear to be an English equivalent. This is the period during which the grape changes appearance: in black varieties the colour starts to show in the skins, and white grapes take on a more translucent appearance, sometimes with hints of yellow in the colour. At this time the metabolism of the grape changes and the sugars start to accumulate at a high rate, pushing down the acid levels simply by dilution.

The right moment for picking depends largely on the balance between sugars and acids being optimal, and this balance will vary according to the style of the wine being produced. In hot climates, where the acid concentration drops at an alarming rate as maturity approaches, picking might take place slightly early. In cool climate conditions, however, the harvest is often left as late as possible, to allow the acids to decrease to an acceptable level.

The graph on the next page shows how the acids have dropped from 17 g/litre to 8 g/litre in twelve days, during which time the sugars have risen from 103 g/litre to 137 g/litre. In this instance, the graph also shows that there is little to be gained by leaving the grapes on the vine after the end of August, as the principal changes have already taken place.

However, this is not the whole story, as what is known as the physiological ripeness of the grape has to be considered. When this has been attained, not only is the sugar-acid balance correct, but the polyphenols will have matured and the pips will have changed from green to brown. Picking too early results in wines that have a green and unripe taste.

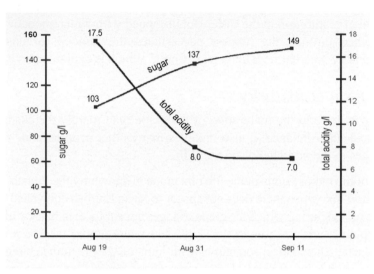

Graph showing the changes that occur during ripening of the grape

Although the use of scientific measurement can be very useful, as seen above, the ultimate test for maturity is to "Go and chew a grape!", and indeed this is what many winemakers do every day as the harvest approaches. In the vineyard, grapes are selected from various positions to give a fair representation of the total situation, and they are chewed, including the skin. If the sweetness and acidity are in balance, the tannins are not too aggressive and the pips are turning brown, then picking can commence.

Chapter 4

THE ROLE OF OXYGEN

Democracies must try to find ways to starve the terrorist and the hijackers of the oxygen of publicity on which they depend.

Margaret Thatcher 1985

The atmosphere in which we live, the air that we breathe, is composed approximately of four-fifths nitrogen (N_2) and one-fifth oxygen (O_2), plus small quantities of other gases. This nitrogen, although critical for life, is relatively unreactive. It requires the efforts of nitrifying bacteria on the roots of leguminous plants to convert it to nitrates and other nitrogen compounds which can then be used by plants to produce amino acids, the building blocks of organic life. The amino acids are then assembled into proteins, a group of substances critical to all life, the name deriving from the Greek *proteios* meaning 'holding first place'.

Oxygen plays an entirely different role, and one that is somewhat ambivalent in that it both supports and destroys life. From the early days of our education, we are taught that oxygen is the staff of life. Oxygen enables fires to burn; oxygen enables our bodies to metabolise food and release energy. Life as we know it could not exist without oxygen.

However, there is another side to the story, because oxygen is also the main element involved in degradation and ageing. Millions of years ago, when life was first emerging, the Earth's atmosphere was based on hydrogen sulphide (H_2S), a gas that is highly toxic to life as we know it today, and oxygen was total anathema. Bacteria are to blame for the change in the atmosphere, for it they who produced oxygen as a result of their metabolism. Ever since primitive life took the route in its development towards an atmosphere containing oxygen, we have had to defend ourselves against its effects.

Oxygen is destructive and needs constant control. Early forms of life had two options: either to retreat to the depths of the oceans, where they could avoid oxygen, or they could learn to defend themselves against it by using it as a source of energy. This is still the situation today, with anaerobic organisms living at great depths in the oceans, while we and other aerobic systems managing to exist in the presence of oxygen, but with the aid of antioxidants. Oxygen is constantly combining with our molecules and destroying them. Iron goes rusty, and we grow old, and fruits lose their flavour. This is all the result of oxidation.

Old-style winemaking

Before the days of scientific training, the traditional winemaker would probably have been unaware of the dangers of oxygen, and would not have taken precautions to prevent its effect. Grapes were pressed in the presence of air; the juice would have picked up plenty of oxygen on its way to the fermentation vessel, and the finished wine would have been moved several times during the clarification process, gathering oxygen at every opportunity. During this time, the fruit flavours in the juice and in the wine would gradually be diminishing and the finished wine would be sadly lacking in 'fruit'.

Obviously, there are many wines that have been made by old-fashioned methods that are superb, or even magnificent. This is because the winemaker has been aware of the dangers of oxygen and has taken sensible precautions to minimise its effect, without actually practising anaerobic winemaking.

Indeed, there are times when aerobic winemaking is totally appropriate, as in the production of some sherries and the wines of Tokaj, where a certain degree of oxidation is essential to the traditional character of the wine. But oxidation has to be kept under control even here, as the pathway to destruction is already being trod, especially if acetic bacteria are present. These are the bacteria that promote the conversion of ethanol into acetic acid by oxidation according to the following reaction:

$$CH_3CH_2OH + O_2 \rightarrow CH_3COOH + H_2O$$
$$\text{ethanol} \quad \text{oxygen} \quad \text{acetic acid} \quad \text{water}$$

This is the very process by which wine vinegar is made, acetic acid being the principle ingredient of vinegar.

Anaerobic winemaking

The principles of anaerobic winemaking originated in the so-called New World countries, where winemaking was approached on a much more scientific basis than in Europe. In Australia, much work was done at Roseworthy Agricultural College and the Australian Wine Research

Institute. The University of California at Davis has also done a great deal of research into modern winemaking techniques.

Modern winemaking is strictly anaerobic, meaning without oxygen. Great care is taken to prevent the ingress of oxygen at all stages because it is known to be the destroyer of fruit in the wine. The prevention of oxidation is not difficult and merely needs good discipline throughout the winemaking process.

The irony here is that wines made by this method, in an all-stainless winery, although very clean and pure, can sometimes suffer what is known as reductive taint. Some of the sulphur dioxide becomes chemically reduced to hydrogen sulphide (H_2S), resulting in the wine smelling dirty.

Antioxidants

The battle against oxygen starts in the vineyard, when the gathered grapes are dusted with an antioxidant powder. This substance is potassium metabisulphite (note the spelling – not metabisulphate), a white powder that is perfectly stable when dry, but which liberates sulphur dioxide (SO_2) when wet. This is most useful at this stage, as any juice that exudes from a berry will immediately be protected from oxidation.

On arrival at the winery, the grapes are crushed and the remaining metabisulphite will dissolve, giving antioxidant protection to the entire mass.

Another antioxidant that is sometimes used is ascorbic acid, otherwise known as vitamin C. It is interesting to note that most of the vitamins are antioxidants, guarding our bodies against the ravages of oxidation.

See chapter 15 for further information on sulphur dioxide and ascorbic acid.

Inert gases

Another important step that must be taken is to prevent the contact of air with the grape juice after it has been pressed out. Flushing all tanks and pipelines with a so-called 'inert' gas such as carbon dioxide, nitrogen or argon removes the air, and thus the oxygen.

In anaerobic winemaking much use is made of what are known as inert gases. In fact, this is a misuse of the word 'inert', which means 'deficient in active properties'. To a chemist, the inert gases (usually referred to as the noble gases) include helium, neon, argon, krypton and xenon, and are truly unreactive and require extreme conditions to make them react with any other substance.

The inert gases used by winemakers are usually nitrogen and carbon dioxide. In this context the word 'inert' is used to mean something nearer to 'harmless', or specifically something that can be used to prevent air from coming into contact with the wine.

Every time the finished wine is moved from vat to vat, or receives any treatment, care is taken to prevent oxygen dissolving. Oxygen dissolved in wine actually reacts quite slowly when left to its own devices, requiring assistance in the form of catalysts to hasten the reaction. Unfortunately, there are plenty of natural catalysts in the form of oxidising enzymes, the oxidases, which are ready to assist. Other catalysts in the form of metal ions also increase the speed of the oxidation process. One of the worst is copper, hence the banning of all copper or bronze equipment in a modern winery (see p.153).

• *Carbon dioxide*

Carbon dioxide is popular because it is cheap and easy to use, being heavier than air. If applied gently at the bottom of a tank, it will displace all of the air. This is an efficient way of removing the oxygen from a tank before filling with must or wine. A simple technique often used to achieve this is to add some blocks of 'dry ice' (frozen carbon dioxide) to a small quantity of wine in the bottom of the vat. The wine warms the dry ice, which causes a rapid evolution of carbon dioxide gas, pushing the air out of the top of the vat. Because of its high density, it is only necessary to use an amount of gas equal to the volume of the tank to remove all of the air.

However, carbon dioxide does have drawbacks. Firstly, it dissolves in wine; the colder the wine, the more it dissolves. Thus, the oxygen barrier is lost, and at the same time the wine gains an unpleasant 'prickle' on the palate. Worse than this, in human terms, is that it is invisible and very dangerous. At a proportion of only five per cent in the atmosphere it can cause unconsciousness and death. Being invisible and heavier than air, many people have died by going down into apparently empty vats to clean them, only to find that they are overcome by carbon dioxide. Death from asphyxiation is rapid, and to an observer outside the vat, the cause of death is not obvious. In the UK, as in many other countries, the laws on entering enclosed spaces are strict: an empty vat should never be entered until the atmosphere inside has been tested and found to be safe, and a second person must always be present outside the vat in case of emergency.

On the positive side, carbon dioxide is an important constituent of many wines, especially white and pink wines. At the correct level (usually around 800mg per litre) it has the effect of making the wine more lively in the mouth. At too high a level the wine becomes 'prickly', and if too low, the wine seems flat and dead. Many producers adjust the level of the carbon dioxide before bottling by sparging (see below) with mixtures of carbon dioxide and nitrogen.

• *Nitrogen*

Another gas that is frequently used is nitrogen, which has different properties from carbon dioxide, the principal one being that it is roughly the same density as air. This is a disadvantage when flushing out tanks, requiring at least three times the volume of the tank to achieve an adequate removal of oxygen. On the other hand, it is almost insoluble in wine, so does not disappear by dissolving in the wine and rendering it 'prickly'. Conversely, it can dissolve slightly in wine which sometimes causes a fine froth to appear on the surface, which does not enhance the appearance of the wine in the glass.

It is much less toxic than carbon dioxide. Being the same density as air, it does not collect in low-lying places and disperses readily. After all, the atmosphere in which we live is about eighty percent nitrogen.

An efficient sequence for oxygen control using inert gases would be:

1. Fill the empty tank with carbon dioxide to flush out all of the air.

2. Fill the tank to the brim with wine, using nitrogen pressure to move the wine through the pipes. (The carbon dioxide is allowed to dissipate from the top of the tank, but see above for the danger.)

3. Allow nitrogen to fill the headspace as wine is removed from the tank, thus preventing the ingress of air.

If pure nitrogen is used extensively during the storage and manipulation of wine, it can result in the wine losing too much of its dissolved carbon dioxide, giving it a somewhat flat taste. This can be prevented by using a mixture of nitrogen and carbon dioxide.

•   *Argon*

For those who do not like the solubility of carbon dioxide and the slightly foaming nature of nitrogen, there is the possibility of using argon, but it is expensive.

One advantage is that its density is close to that of carbon dioxide, so it is more efficient at purging the air from a vat than nitrogen. It is also very good at removing dissolved oxygen from the wine by sparging, (see opposite) but care is needed as it is easy to strip the flavour elements from the wine. Apart from being expensive, it is not always easy to find a source of food-grade gas.

Dissolved oxygen

Oxygen possesses a dual nature: in the right place at the right time, it aids the development of wine, but when allowed unlimited access, it can lead to its destruction. Oxygen dissolves readily in water, as anyone knows who keeps fish. Since wine is at least 85% water, it is not surprising that oxygen will dissolve in wine if it is allowed contact with air. The principle defence against the ravages of oxygen is sulphur dioxide, but every molecule of oxygen that dissolves in wine destroys four times its weight of sulphur dioxide, as shown by their relative molecular weights:

$$2SO_2 \quad + \quad O_2 \quad \rightarrow \quad 2SO_3$$
$$128 \qquad\qquad 32$$

Dissolved oxygen (DO) is eventually destroyed by the sulphur dioxide, but this does not happen immediately, research having shown that dissolved oxygen and sulphur dioxide can coexist in wine before inter-reaction takes place. During this time every molecule of dissolved oxygen is a threat to the quality of the wine, taking it another small step towards destruction. The only secure way of preventing oxidation is by preventing air from coming into contact with the wine by the use of inert gases and by keeping vats brim-full and sealed.

The importance of the control of dissolved oxygen has been somewhat neglected, even in the recent past. Many winemakers are still totally unaware of the critical nature of this aspect of quality control, partly because the older generation were not trained in this work, and the measurement of dissolved oxygen has been ignored. Modern meters will give accurate results, provided the operator understands the pitfalls that come with bad sampling and the manner in which measurements are made. The atmosphere contains 20% oxygen, so all measurements have to be made anaerobically if they are to have any meaning. Every time wine is manipulated, a new measurement has to be made, and if there is an increase in the DO, steps have to be taken to investigate the reason, and modifications made to the procedure or process.

Sparging

This is the ultimate refinement in the use of inert gases, and is the process of injecting fine bubbles of gas into a liquid, usually to remove dissolved oxygen. As the bubbles pass through the liquid, a complex interchange takes place between volatile elements in the liquid and in the gas. If nitrogen is used as the sparging gas, any oxygen dissolved in the wine passes into the nitrogen bubbles and is thereby removed.

Unfortunately, nitrogen will also remove carbon dioxide, making the wine taste flat. Mixtures of carbon dioxide and nitrogen of varying proportions can be used during the sparging process to control the level of this gas. A high proportion of carbon dioxide in the sparging gas will increase the level in the wine; a low proportion will reduce it.

The danger with sparging, as with so many other wine treatments, is that it can easily be over-used. The removal of dissolved oxygen is an excellent principle, but what must be borne in mind is that sparging does not remove oxygen exclusively. It will remove anything volatile, and flavour components are by their very nature volatile. So care must be taken to monitor the level of dissolved oxygen, and to use the sparging process sparingly.

The positive role of oxygen

With the advent of modern methods of anaerobic winemaking we tend to think of oxygen as the great destroyer of all that is good, but this would be wrong. Oxygen is not totally destructive, and there are several stages in the winemaking process where it plays an important role:

1. Sometimes the must is given a short burst of oxidation during the clarification process to rid it of the more fragile components so that the resulting wine has a longer shelf life. This is a process known as hyperoxidation (see p.62).

2. Before the fermentation starts, it is good practice to saturate the must with oxygen, because this gives the yeast a 'kick start' into action. When yeasts are in an oxygen-rich environment they reproduce more rapidly and can therefore build up a large population quickly, which results in a prompt start to the fermentation. (See chapter 7)

3. If a fermentation is slowing down as a result of the yeast population being too low, a quick shot of oxygen revitalises the yeast. This is easily done by pumping the fermenting must over a cascade (see p.83).

4. During its period in wooden barrels a tough red wine becomes silkier and softer because the oxygen that permeates the wood oxidises the harsh polyphenols which drop out as a black deposit in the barrel. Sometimes this process is achieved by adding oxygen to the wine in a tank instead of putting the wine into a barrel. This process is known as micro-oxygenation (see p.133).

5. It has been realised that wine needs a small quantity of oxygen after bottling during what has been known as the anaerobic maturation period in order to prevent it from getting a reductive taint. This is when the sulphur dioxide becomes reduced to hydrogen sulphide because of a lack of oxygen, as shown by the use of screwcaps which give a seal that is too perfect (see p. 206).

6. Wines made by the anaerobic method are superbly clean and fresh, and full of youthful fruit, yet risk being boring by being too perfect. This is another area where the art of the winemaker comes into play, by knowing how to apply the many other means of adding character and preventing the wine tasting like the product of a wine 'factory'. A controlled amount of oxidation undoubtedly adds to the complexity and character of the wine.

Chapter 5

PRODUCING THE MUST

No grape that's kindly ripe, could be
So round, so plump, so soft as she,
Nor half so full of juice.

Ballad: Upon a Wedding
Sir John Suckling 1609 - 41

The product of crushing is known as 'must', from the Latin *mustum*, meaning new or fresh, which is somewhat ironic, considering another word in the English language with a wider usage is 'musty', which means exactly the opposite. It should be noted that must is not just the juice but, in the case of red wine production, it is juice plus skins.

In the days before the introduction of scientific principles to the art of winemaking, grapes were harvested when they tasted sweet and fleshy, and they were crushed without undue delay by the action of the human foot. Sometimes this crushing was achieved, as in the production of port, by the naked foot, which proved to be ideal for the purpose, being firm yet gentle in action. In other areas, such as Jerez, special boots were worn, with nails projecting from the sole, to prevent the pips from being crushed and thereby releasing their bitter contents. These old practices have not entirely died out, but their use tends to be more for marketing than for practical reasons (but see p.119).

Pressing was always carried out in a traditional basket press, because this was the only design available, with all of its attendant problems. The tedious and slow manipulation was of no consequence, there being ample labour and time in plentiful supply. These conditions were fine for a small vineyard owned by one family, producing enough wine to make a decent living.

With increased volumes and the greater influence of accountants and scientists, the situation had to change. Developments in machinery meant better and quicker crushing and pressing operations. New processes, such as flotation and hyperoxidation, were introduced. The modern winemaker has a vast repertoire of techniques from which to choose, with the possibility of producing better wine, and more cheaply than ever before.

However, one general principle remains, that at all stages in the process skins should be treated gently. The more abrasion that the skins receive, the harsher will be the juice. This is due to the fact that the skin cells contain tough polyphenols that will be released into the juice if the cells are ruptured.

Harvesting the grapes

As the time for harvesting draws near, a major decision has to be made regarding the method of picking. This is a comparatively easy decision, for there are only the two alternatives, picking by hand or by machine. The parameters to be considered are based on a combination of quality considerations, speed, economics and feasibility.

• Picking by hand

Hand harvesting is essential where the selection of the finest quality grapes is paramount. The great sweet wines, such as Sauternes, Tokaji Aszu or Beerenauslesen depend upon the selection of individual noble-rotted berries. These vineyards have to be traversed several times over to gather the grapes one by one, when the action of *botrytis cinerea* has shrivelled the grapes to their peak of condition. This selection (Fr. *triage)* is sometimes done in the vineyard, but can also be carried out at the winery when the bunches are sorted according to quality on a *triage* table (see p.45).

Hand picking
Aglianico grapes
in Basilicata

Human labour is again called upon if whole bunches are required, e.g. for carbonic maceration (see p.97), or when the stalks are needed for added tannin.

A further consideration is the arrangement of the plants in the vineyard. There are many older sites where it would be impossible for a machine to function: bush vines planted randomly, rows that are too close, grapes hanging at different levels and concealed under the canopy.

The disadvantage of this traditional way of gathering in the harvest is the length of time it takes, even with a large band of pickers – and it is expensive, being some ten times the cost of machine harvesting.

• *Machine harvesting*

The machine harvester comes into its own as the size of the operation increases, but the greatest advantage of the machine is its speed, ensuring the gathering of the grapes when they are in their peak of condition.

Machine harvesting at Chateau Tariquet

Using one of these machines, the winemaker can choose the exact moment of picking, instead of having to spread the operation over several days, during which time it might rain or excessive sunshine might take the grapes beyond the optimum ripeness.

One of the great assets of the machine is harvesting during the night. This is particularly helpful in hot climates, when the grapes can be gathered at their coolest, thus minimising deterioration and the cost of cooling the juice before fermentation. Night harvesting has almost become the norm in hot climates and large vineyards, as are found in Australia and California, where grapes are often transported hundreds of miles to the winery.

There are, however, constraints on the use of machine harvesters because these monsters can only operate in specially planted vineyards and on relatively flat land. Nevertheless, at least one renowned producer of Chablis Grand Cru uses a machine in preference to hand picking because he can optimise the picking to the nearest day.

Grapes ready for harvesting

The machine picks the grapes, leaving the stems on the vine

The grapes are gathered by shaking the vines, which causes them to fall off, leaving the stems attached to the plant. This ripping action tears the skins and exposes some of the pulp to the atmosphere, so protection against oxidation is important. There can be no selection of grapes, and there are no stalks as these will have been left on the vine. Neither can the grapes be used for the carbonic maceration process (see p.97), because whole bunches of undamaged grapes are required for this purpose.

Interior of harvester showing the vibrating bars and the grape conveyor

That having been said, machine harvesting is widespread and is gaining in popularity, because its advantages of speed and convenience outweigh the shortcomings. Despite the denial by some of the owners of classed growth chateaux regarding the use of mechanical harvesters in Bordeaux, over 700 such machines have been counted in operation in that appellation. In common with many other techniques, the achievement of good quality comes not from the technique itself, but from the way it is handled.

Harvesters have been greatly improved over the last few years. The vicious metal bars which used to shake the vines have been replaced by curved plastic ones that are much more gentle and cause less damage. Also, a modification of a de-stalking machine is now installed over each grape hopper, so that all foliage and pieces of vine are separated from the grapes. Winemakers refer to this extraneous matter as MOG (material other than grapes). There can be no doubt that machine harvesting will be the preferred method in the future.

Transport to the winery

Whole bunches of grapes are relatively stable when first picked and can be kept for a while before processing (see p.104). But grapes are often damaged during the harvesting process, releasing some of the juice, which is not stable and is readily oxidised by the oxygen in the atmosphere, especially as the oxidising enzymes are active. Traditionally, pressing used to take place as soon as possible, hence the building of wineries in the middle of vineyards. Even in this situation, preventative measures have to be taken to avoid loss of quality by oxidation. It is a common sight to see grapes being dusted with potassium metabisulphite to knock out the enzymes and slow down the oxidation (see p.33).

With larger operations grapes have to be taken some distance to the winery. Where truly anaerobic winemaking is required, as in the production of modern, fruit-driven wines, quite elaborate means of transport have been devised. The chances are that the grapes will have been harvested by machine and have therefore been de-stemmed and exist as separate berries. This makes it possible to transport them in enclosed tankers rather than open trailers.

Grapes being loaded into a tanker from the trailer.
The white box contains dry ice.

Grapes being discharged into the receival bin at the winery under a blanket of carbon dioxide.

Taking anaerobic principles one step further, it is relatively easy to blanket the grapes with carbon dioxide gas as they are loaded into the tanker, and again as they are discharged into the receival bin.

Sorting

The best wine is made from the best grapes, and for this reason the truly quality-conscious winemaker needs to sort the good grapes from the bad (Fr. *triage*). This can be done by passing the bunches of grapes along a moving belt, known as a triage table, with people each side trained to select the best grapes for the *grand vin*, while the poorer grapes go into the lesser wine or are possibly discarded altogether.

Grapes being sorted for the production of Tokaji wine.

De-stemming

A major decision point has been reached: to remove the berries from the stems or to leave the bunches whole. This decision is based largely upon tannin control, with ease of processing as a secondary factor.

Stems are a rich source of tannins and can be used to raise tannin levels by including them in the maceration process. If there is already a sufficiency of tannins in the grapes, the stems from all or part of the crop can be removed by a de-stemming machine (Fr. *égrappoir*). However, this does result in a more difficult pressing operation because the stems act as useful drainage channels. A technique sometimes used to improve drainage is to de-stem the whole crop and then to pack the press with alternate layers of grapes and stems, which produces an easily drained mass.

The de-stemming machine is simple in design and operation, consisting of a rotating, perforated drum, with contra-rotating blades inside the drum. The bunches of grapes are fed into the rotating drum, where the blades tumble the bunches vigorously, causing the berries to fall off and ultimately to fall out through the holes, leaving the stems inside the drum. The berries fall on to a conveyor belt which transports them to the winery, and the stems are collected as they pass out through the end of the drum.

The stems that have been removed are sometimes simply dried and burned as fuel, but more ecologically-minded winemakers have

Traditional de-stemming machine with rotating vanes

discovered that stems contain sufficient sugars to make it worthwhile to subject them to further processing. The stems are chopped and soaked in water to extract the sugars, which are then fermented to produce a weakly alcoholic liquor. This is subsequently distilled to give a high quality spirit of vegetable origin and, as such, can be used for fortification purposes. (Alcohol of mineral origin, such as is manufactured indirectly from petroleum, is not permitted for use in foodstuffs.)

Crushing the grapes

Another decision point has been reached: to crush the grapes or to leave them intact for direct pressing or even for fermentation (see p.100).

It is worth noting that crushing and pressing are quite distinct processes, used for two different purposes. Crushing amounts to nothing more than ripping the grape apart to allow the free-run juice to flow out. The rollers in the crusher are not in contact with each other and pressure plays no part in the process. Pressing, on the other hand, is the extraction of the remaining juice which is contained in tougher cells situated nearer to the skin of the grape, and can only be released by rupturing the cells under pressure.

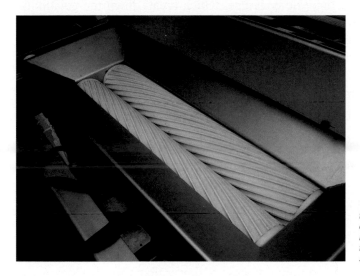

The rollers of the crusher do not meet but merely tear the grape skins apart

The romantic image of crushing is of a group of men, with legs bared, dancing around in a tub of grapes. Although this technique is still sometimes used for the production of port (see p.119), nowadays mechanised methods of crushing are used to increase the speed of the operation. Although deceptively simple, the crusher (Fr. *fouloir)* incorporates considerable experience in its design. The crushing process must be carefully managed, as the objective is to release the free-run juice and to reduce the solid parts of the grape to the correct condition for fermentation and maceration, whilst avoiding damage to the pips. The skins must not be torn away from the pulp because this affects the extraction of aromas from the skin cells. The stalks, if present, must not be torn open because too much harsh tannin would be released. For these reasons, the rotating parts of crushing machines are carefully designed and have an adjustable gap between them of around three millimetres.

The free-run juice, which is held in large and somewhat fragile cells occupying most of the volume of the grape, is liberated. This juice is of high quality and contains the lowest tannin level, because very few skin cells are fractured during the crushing process.

A comparatively new technique for releasing more juice from the pulp is the use of pectinolytic enzymes, which break down the gelatinous pectins in the pulp, thus releasing more juice from the sticky mass. However, some winemakers will not use them at this stage, maintaining that there is a reduction in the characteristic flavour of the variety.

Draining the juice

The manner of the separation of the free-run juice is important in the making of white wine, because it has an effect on the quality of the wine. The secret of good draining is to minimise the disturbance of the skins to prevent the extraction of harsh polyphenols. When making white wine in a small scale winery the crushed grapes go straight to the press, but with larger volumes it is necessary to use dedicated draining equipment. The best method is the static drainer, where the juice can percolate gently through the mass of skins without any disturbance. However, this method can pose a problem of through-put in a large winery because the draining process is slower than the other stages.

Mechanical drainers have been introduced, consisting simply of a rotating archimedes screw in a horizontal tank with a perforated base. The screw rotates once every two or three minutes, gradually pushing the grapes through the machine, allowing them to drain as they go. Whilst overcoming the lack of speed of the static drainer, the disturbance of the skins does result in a greater release of the polyphenolic substances.

In the production of white wine the yield of free-run juice is typically between 400 and 500 litres per tonne of grapes.

Pressing the berries

In the making of white wine, pressing comes next because it is necessary to release the remainder of the juice (usually about 150 to 200 litres per tonne of grapes), which is held in the smaller, stronger cells which make up the inner layer of the skin. At the same time more polyphenols and aroma compounds are released.

The quality of expressed juice is usually in inverse proportion to the degree of pressure used. The first juice extracted with the minimum of pressure is of good quality, although containing a higher proportion of polyphenols. This is often exactly what the winemaker needs to produce the correct balance in the final blend. It is only the later fractions, pressed out with greater force, that are of progressively lower quality, with the final fractions used only for vinegar production or for distillation. The greater the pressure, the more of the skin cells that are broken and the greater the concentration of the unpleasant, bitter tannins. Thus, the principle in good pressing is the use of minimum pressure to achieve the necessary extraction of juice.

In the production of red wine, this part of the process comes after fermentation, but the principles of pressing are the same for the production of both white and red wines.

• The basket press

When the pressing of grapes was first mechanised, it was the basket press that performed the task. Otherwise known as the vertical screw press, it has remained virtually unchanged for over a thousand years. The mode of operation is simple, merely increasing the pressure on the mass of skins by screwing down the lid of the press, causing the cells to rupture and releasing their contents. The great drawback is that the process cannot be hurried. Increasing the pressure too rapidly results in a broken press.

The problem lies in the physical nature of the solid mass, which is gelatinous and sticky, and out of which the juice can only exude slowly. To make things worse, the juice trickles out between the wooden slats of the basket, runs into a trough around the edge and from thence into a tub, all the time dissolving oxygen from the atmosphere. At the end of each pressing, the press has to be unwound, the skins cleared out with shovels and the press re-filled, all of which are labour intensive operations.

Modern basket presses with hydraulic control

Despite these facts, the basket press is still used by some very high-class wineries in preference to more modern presses because the static bed of skins acts as a fine filter, yielding a fine juice of good clarity. Chateau Petrus, for example, uses this style of press. What more need be said? Most of the presses used for champagne production are based on the basket press, using a basket that is shallow and of large diameter to enable the juice to get away from the skins as quickly as possible, thus minimising the extraction of colour from the skins of the black grapes.

• *Horizontal screw press*
The horizontal screw press was the first stage in the evolution of presses as typified by the Vaslin press, which is effectively a basket press turned through ninety degrees and given two pistons, one at each end

of the slatted cylinder, with an access port on the side of the cylinder. The cylinder rotates on a horizontal axis about a stainless steel screw, on which are threaded the two pistons.

The ingenious element of the design is that the screw is divided into two halves, one half being a right-handed thread, the other left-handed. The result is that if the screw is held stationary, when the press rotates, the two pistons move towards each other, the reverse rotation winding the pistons apart. Stainless chains are connected between the pistons, and as the chains straighten they break up the mass of skins after each pressing. Filling and emptying is simple, and the operation can be totally automated, even to the extent of automatic delivery of the juice from each pressing to separate vats, making it much more efficient in terms of time and labour.

Interior of a Vaslin press showing the thread, chains and piston

The complete sequence of operation is:

- With the door at the top, the press is filled and the door closed.

- The whole press is rotated, including the screw thread. This tumbles the grapes and releases the maximum free-run juice.

- While the rotation continues, a brake is applied to the screw thread to stop it rotating, which causes the pistons to move towards each other.

- Rotation continues until the required pressure has been reached.

- The rotation is reversed, to part the pistons, the chains breaking up the mass of skins.

- Pressing is repeated several times, using a higher pressure each time.

- At the end of the operation, the door is opened and the press rotated until the door is at the bottom, when the skins fall out and the press can be re-filled for the next batch.

Vaslin presses in the bodega of a large sherry producer

Its only shortcoming is that it produces a rather coarse juice due to the abrasion of the skins caused by the tumbling, and the high pressures that are required as the skins are progressively compressed into a smaller volume in the centre of the cylinder. This reduces the effective surface of the press to that portion between the two pistons. It is not unknown for pressures in a Vaslin press to reach 30 bar, which is 30.6 kilograms per square centimetre or 435 pounds per square inch!

• *Pneumatic press*

The Willmes company developed the pneumatic press in order to overcome the problem inherent in the design of the horizontal screw press, so that a more efficient pressing could be achieved and at lower pressures.

Willmes press showing the black bag that expands and presses the grapes

In this press, the two pistons are replaced by a cylindrical pneumatic bag, or bladder, or sausage, which can be inflated with compressed air (or even cold water, if cooling is required). As the bag expands, the skins are pressed in a thin layer against the entire surface of the slatted cylinder, enabling the juice to exude under very low pressures, as low as 0.1 bar, resulting in superior quality.

The common fault with all of these presses is the danger of dissolving oxygen from the atmosphere, with the risk of destroying some of the more delicate constituents.

• *Tank press*

The tank press is a development of the pneumatic press, where everything takes place inside a closed tank, which can be pre-flushed with nitrogen and the juice pumped away without ever touching air. At present, this design is regarded as the ultimate by modern anaerobic winemakers because the combination of low pressures and gas flushing gives a high quality juice. Unfortunately, complex equipment is inherently expensive.

In this press the rubber sausage is replaced by a flexible diaphragm which divides the cylinder horizontally into two halves. During the filling, the diaphragm lies in the bottom of the press. To start the pressing process, air is pumped into the space underneath the diaphragm, which then rises up and squeezes the skins against the collecting channels in the top half of the cylinder. If true anaerobic working is required, the press can be flushed with an inert gas before being filled with grapes, and the juice can be pumped out of the press into a tank that has also been flushed with inert gas.

Interior of a tank press showing flexible juice collecting channels

All of the presses described above suffer from the same shortcoming: they are batch processes. They have to be loaded with grapes, operated to extract the juice and then emptied; a sequence which is tedious, time-consuming and labour intensive. Wineries handling very large quantities of grapes need presses that can work on a continuous principle, rather than the batch system of all other designs.

• *Continuous screw press*

The continuous screw press was developed for large wineries producing huge volumes of wine of commercial quality. It achieved ill repute in its early days by virtue of the fact that it can squeeze grapes almost bone-dry, yielding foul juice fit only for brandy production.

The original continuous screw press consists of a strong steel tube, perforated along its length, containing an archimedes screw, which pushes the grapes through the pipe. Pressure is achieved by using a variable pitch screw, the turns coming closer together towards the outlet end of the press. The result is that the volume between each turn of the screw becomes steadily less as the grapes go through the machine. This generates huge pressures and produces juice of dubious quality.

Interior of a continuous screw press showing the strongly-built slatted tube

However, by a clever modification it is possible to take juice of a very acceptable standard from the low-pressure end of the press. The simple change to the design incorporates a series of adjustable troughs underneath the press which catch juice escaping from the different sections of the tube, each fraction being diverted to a different vat.

Another version of the continuous screw press uses a simple archimedes screw of constant pitch, which turns to introduce grapes into a chamber at the far end of the screw. The whole screw then moves forward to press the skins against the end of the chamber. The screw then moves back, still turning, to introduce more skins, which are pressed on the next forward stroke of the screw. The advantage of this mechanism is that the pressure can easily be regulated by adjusting the forward action of the screw. Both of these modifications of the continuous press can be found in use in large modern wineries, where wine of good commercial standard is produced in large volumes.

Yet another variant is to feed the grapes into the press with a normal archimedes screw, with the far end of the press closed by a strong steel door that is held closed by an hydraulic piston. The force that this piston exerts on the door can be adjusted so that the door opens when the pressure inside the tube reaches a pre-defined level. If set to a low pressure, this style of press can produce good results.

Many winemakers reserve the use of the continuous press for the production of must which is intended for distillation into brandy, and in some countries, such as Algeria, its use is banned for the production of wine of controlled origin.

Chapter 6

ADJUSTING THE MUST

The real world is not easy to live in. It is rough; it is slippery. Without the most clear-eyed adjustments we fall and get crushed. A man must stay sober: not always, but most of the time.

Clarence Shepard Day 1874-1935

Throughout the winemaking world oenologists have been tempted to meddle with the natural balance of the must when the seasonal weather has been less than perfect. Despite the fact that it can be very frustrating at times, nature has a remarkable way of knowing what is best. Man has discovered that he can adjust what nature has produced, but only to a limited degree: the further we move away from the natural balance, the worse becomes the finished wine. Thus, in a poor year, when the normal acid-sugar balance is upset, small adjustments will yield an improvement, but larger changes make matters worse. Too much added acidity makes an 'angular' wine, and large volumes of added sugar merely dilute the natural fruit of the grape. It is quite logical, therefore, that wine producing countries have regulations that limit the adjustments that can be made to juice and wine, although these differ from country to country.

The only changes to the natural balance that are allowed are the addition of sugar and the addition or removal of acid, and only when specifically stated in the regulations, and within clearly specified limits.

An interesting principle that has been noticed from experience is that any adjustments that are felt necessary are best made to the must before fermentation, rather than to the finished wine. It would appear that the process of fermentation helps the adjustments to 'marry' more thoroughly, resulting in a wine that is more harmonious.

At this stage the must is in a precarious state, prone to attack by oxygen and by a host of different microorganisms. It is a rich and nutritious medium and must be protected until the moment of fermentation, when all winemakers heave a sigh of relief because the active yeast and the copious production of carbon dioxide combine to give the must considerable protection. Until this moment, recourse is usually made to the protective action of sulphur dioxide.

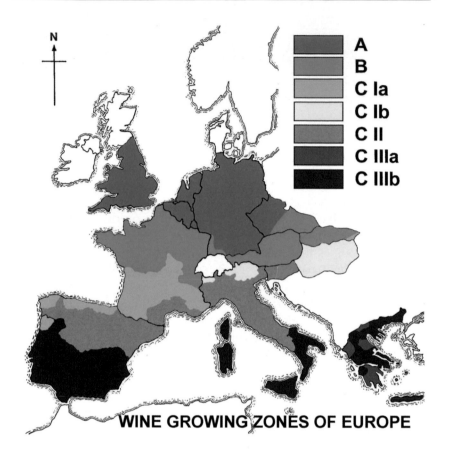

WINE GROWING ZONES OF EUROPE

In order to administer the European regulations regarding aspects of grape production and wine making, Europe has been divided into wine growing zones that essentially follow the climatic differences across Europe. Zone A is the northernmost belt, covering England, Wales and most of Germany, zone B includes northern France and the southernmost part of Germany, while zone C gathers up the rest of Europe. As one has come to expect of modern bureaucracy, three simple zones did not suffice for the administration of complex regulations, so the zones have been sub- and sub-sub-divided! This enables the limitations on sugar and acid adjustment to be applied differently to quite small areas of Europe.

Sulphur dioxide

The initial addition to the grapes in the vineyard is made on a rather crude basis, but when the grapes have been converted into must, the chemist can make a proper analysis and the must can be adjusted to an appropriate level.

The use of sulphur dioxide at this stage is critical, especially in the production of white and pink wines, because the next stage is clarification, which takes two or three days for completion. Unfermented must is in a delicate state: it can be subject to oxidation and also to premature fermentation. Sulphur dioxide protects against both of these troubles. (For more detail on sulphur dioxide, see p.165.)

In white wine production, the use of sulphur dioxide is universal during the clarification process (unless the hyperoxidation technique is being used, see p.62). It protects the must against oxidation, kills the bacteria that could degrade the must and stuns the weaker yeasts, thereby giving the good yeasts a chance to develop.

In red wine production, the action of sulphur dioxide is somewhat different. Red musts contain the skins, which are covered with a high population of microorganisms that cannot be totally controlled by the sulphur dioxide. Some of these organisms become active, producing acetaldehyde which binds the sulphur dioxide as bisulphite addition products, thus removing all of the free SO_2. To make matters worse, sulphur dioxide also combines readily with some of the anthocyanins, so one might come to the conclusion that the use of SO_2 in red wine making is pointless.

Many winemakers, however, have observed that the use of sulphur dioxide improves the extraction of polyphenols from the skins by a mechanism that is not fully understood. So, despite the drawbacks, sulphur dioxide is used at this stage for both white and red musts.

It is possible to make wine without the use of sulphur dioxide, but extreme precautions have to be taken at all stages to prevent the ingress of oxygen because the wine will be very susceptible to oxidation. It is interesting to note that its use is permitted even in organic winemaking (more correctly, wine made from organically grown grapes).

Clarification (white and pink wines)

Freshly pressed must contains considerable quantities of solid matter, mostly cellular debris from the skins, which can cause off-flavours in the wine if not removed. Thus, clarification has become the norm in most wine producing areas of the world.

However, the degree of clarification is very much part of the expertise of the winemaker. There is a danger in getting too enthusiastic about clarification, the solid particles being nutritious to yeasts because they have adsorbed on to their surface amino acids, minerals and vitamins. (Note that *ad*sorbed indicates a surface attraction, contrasting with *ab*sorbed, meaning distributed within the body of the solid, rather like a sponge.) The complete removal of these particles can reduce the nutritional value of the must to such a degree that the yeast has difficulty in starting a fermentation. It is a matter of knowing the grapes and the must, knowing how rich they are in nutrients, and knowing how much clarification is necessary.

• *Settling*

Settling (Fr. *débourbage*) is the traditional method used around the world, where the must is left to stand for 12 to 24 hours or even longer, during which time the particles sink to the bottom of the vat. This is a rather long period for must to be left in an unprotected state, but gravity is not a strong force and the particles are not very dense or large, and fall somewhat slowly. The protection given by sulphur dioxide during this time is critical, because it not only acts as an antioxidant but also reduces the activity of yeasts and bacteria. For the same reason, the must is often cooled to below 15°C, which slows down both the bacterial activity and any oxidation. At the same time, the cool wine will dissolve oxygen more readily, a factor which is favourable to the rapid start of fermentation (see p.74).

• *Centrifuging*

A centrifuge is an expensive machine that is based on a rotating conical drum that spins at very high speeds and can separate solids from liquids by centrifugal force. Centrifuging is a modern method used in large wineries to shorten the period of clarification considerably, as a large vat can be clarified in one to two hours.

Centrifuging can be applied at any stage in winemaking where clarification is needed, e.g. before fermentation, after fermentation to remove yeast cells, or after fining to remove the deposit. However, some winemakers regard the centrifuge as a harsh tool that can cause damage to wine or must, so caution is necessary.

*Centrifuge is useful
in large wineries
for rapid clarification*

One of the great dangers of using a centrifuge is that it can saturate the liquid with oxygen from the atmosphere, as the liquid is spinning in a thin film inside the cone. Although this could be an advantage at the must clarification stage, at any other stage it is necessary to flush the interior of the centrifuge with nitrogen before use, to prevent the oxygen dissolving.

• *Flotation*

Flotation is another process which is sometimes used for clarification and has been in use for decades in the mining industry for separating finely divided particles of ore from water. Small bubbles of nitrogen are blown up through the must, catching the solid particles and floating them to the surface, where they can be skimmed off by a rotary suction device. This can be adapted for the dual purpose of clarification and hyperoxidation at the same time by using air rather than nitrogen for the bubbling process (see p.62).

Hyperoxidation

It is widely accepted that wine after fermentation must be protected from the ravages of oxygen if it is to remain fresh and with a good fruit flavour. Opinions differ, however, regarding the oxidation of must before fermentation. Some winemakers believe that must that has been too well protected from oxygen results in a wine that is more sensitive to oxidation. Conversely, must that has been allowed some contact with oxygen yields a more stable wine. This is the basis of the process known as hyperoxidation.

Hyperoxidation is a somewhat surprising technique that involves the deliberate oxidation of the must before fermentation, something which might be regarded as anathema to a modern winemaker. During this process, oxidation is allowed to take place, thus destroying the more susceptible components of the juice, resulting in a finished wine that is more stable towards oxygen.

It is carried out by putting the juice into a specially adapted tank with a series of inlets at the bottom through which air can be bubbled. Additions of sulphur dioxide must not be made as this would prevent the deliberate oxidation that is required. The first stage is simply the clarification of the juice by flotation, the solid matter being skimmed off the top of the liquid.

The second stage of the process is the oxidation of the juice, during which time it turns the colour of black coffee, which must be rather frightening to the winemaker when seen for the first time! During this period, the various components of the wine are gradually oxidised, starting with the most fragile and gradually working towards the more stable compounds. The skill in this technique is knowing when to stop the oxidation and to commence fermentation. Hyperoxidation is not a process that should be adopted casually in the expectation of producing a highly stable, fruity wine at the first attempt. Too much oxidation results in a destruction of the fruit which is at the heart of the wine.

Surprisingly, during fermentation some of the oxidation of the must is reversed, and the colour returns to normal. This effect is known as reduction, which is a chemical term that means the opposite of oxidation: oxidation is the addition of oxygen, reduction is the removal of oxygen. Fermentation is a reductive process and can reverse the oxidation, but this reversal of oxidation is not total, and only the most fragile components of the juice are permanently destroyed, hence the improved stability of the finished wine.

Results from the use of this technique have confirmed that wines made by this method are indeed more stable and have a longer life. A large producer of wines in the south of France has virtually doubled the shelf-life of its delicate rosé wine since adopting this technique.

Acidification

Grapes grown in hot climates, such as southern Europe, Australia or California, will possibly be lacking in acidity. This is easily rectified by adding acid, usually tartaric, which is the natural acid of grapes.

Citric acid, the acid of citrus fruits, is cheaper and readily available, but can be metabolised by some yeasts and converted into acetic acid, resulting in an increase in volatile acidity. In Europe, only tartaric acid is allowed and indeed, acidification itself is only allowed in the hottest parts of Europe, zones Cii and Ciii, and only by a maximum of 1.5 g/l expressed as tartaric acid.

The process of acidification is very simple; all that is involved is an initial tasting and analysis, followed by a calculation of the required amount of acid to be added. The acid is weighed out, dissolved in a little juice, added to the bulk and stirred.

The old Jerez practice of plastering, as it was known, involved the addition of calcium sulphate (plaster or gypsum) to the must or wine. This increased the acidity by precipitating calcium tartrate and leaving the more acidic potassium bisulphate in solution. This practice has virtually died out as it is more effective simply to add tartaric acid, as with any other wine.

Although adjustment of acidity is best done before fermentation, giving a more integrated wine, the regulations allow for this to be done both before and after fermentation, although the limitations on adjustment in the finished wine are tighter.

Deacidification

In cool climates, which in reality prevail in the greater part of Europe, sunshine is usually in deficit, with the result that the acid level even in ripe grapes is on the high side. Hence, European regulations permit deacidification anywhere in Europe with the exception of zone Ciii(b), which is the hottest zone and is usually desperate for more acid, not less!

Although chemically simple, the mechanism involved in taking acid out is not as straightforward as adding it. This is because acid cannot be physically removed, but has to be neutralised by a chemical reaction within the must itself. Furthermore, the acidity is not caused by a single acid, but is due to a mixture of acids, mostly tartaric and malic (see p.23).

The reduction in acidity, or deacidification, is achieved by the addition of a carbonate such as calcium carbonate (chalk) or potassium bicarbonate, which neutralises the acids in the must. This is precisely the same reaction as the old-fashioned remedy for indigestion, when bicarbonate of soda is taken to reduce the acidity of the stomach.

The addition of chalk removes mainly tartaric acid, the product of the reaction being calcium tartrate, which comes out as calcium tartrate crystals. However, this causes great problems with tartrate deposits in the resultant wine because calcium tartrate forms crystals only very slowly, even with chilling, because its solubility remains much the same at all temperatures.

Many winemakers prefer to use potassium bicarbonate because it does not involve the addition of calcium ions, thus avoiding the complications of the difficult removal of calcium tartrate. Potassium bitartrate crystallises relatively easily because its solubility decreases considerably with the lowering of temperature.

These procedures, however, only remove the tartaric acid as tartrates; there is little effect on the malic acid because its calcium and potassium salts are very soluble and do not precipitate out. Grapes grown in cool climates will have an abundance of this acid, giving such wines a particularly sharp taste.

A technique sometimes adopted is to move a portion of the wine into a different vessel and then to add sufficient carbonate to neutralise all the acidity. The neutralised wine is then blended back into the bulk. This technique proportionately lowers the acidity due to all the acids, rather than that due only to tartaric.

•   *Acidex*

Another practice is the use of what is known as double-salt deacidification, which uses specially prepared calcium carbonate known by the brand name of Acidex. However, because of the expense, its use has been limited to those wines that have excess malic acidity and have been grown in cool climates such as England or Germany.

This product contains a small proportion of finely powdered calcium tartrate-malate, a complex double salt that eliminates both tartaric and malic acids by forming crystals of calcium tartrate-malate which is very insoluble. But its use is not straightforward because it depends upon the presence of malic and tartaric acids in roughly equal proportions. It is necessary, therefore, to analyse the must for both acids before treatment. In a cold climate it is highly likely that the malic acid will be more abundant than the tartaric. So the strange situation will arise that extra tartaric acid will have to be added in order to remove the malic acid and therefore achieve an overall lowering of the acidity.

Enrichment

The practice of adding various forms of sweetening agents to must prior to fermentation appears to have started during the last quarter of the eighteenth century, although it was not realised at the time that the improvement in quality was due to an increased alcoholic content.

It was not until 1815, at the end of the Napoleonic wars, that the addition of beet sugar became officially recognised. Jean-Antoine Chaptal, Comte de Chanteloup, once Minister of the Interior and Treasurer of the Senate, and a chemist by profession, suggested that the excess production of beet sugar could be used for enriching wine. Thus was born the practice still known as chaptalisation, a dubious memorial ill-befitting a man of great talent. Although the name of Chaptal is still associated with this process, it is gradually being dropped, particularly by the French, because of the ill repute that the misuse of this process has brought to French winemaking.

Enrichment has probably caused more controversy and more subterfuge than any other treatment. It is closely regulated in all European countries, according to the laws of each country. It is not allowed in Italy or Spain; in France its use varies from region to region, and, in Burgundy, from district to district.

Under cool conditions the natural sugar content of the grape is not sufficient to give an adequate level of alcohol in the finished wine. In these circumstances, sugar is added to the must before fermentation, to boost the alcohol level in the finished wine. But what is an adequate alcohol level? Opinions differ widely. One fact is indisputable: the more sugar that is added to the must, the more the flavour is reduced, by the simple laws of dilution. On the other hand, if the alcohol is too low, the balance of the wine is affected. The correct level of enrichment is one of the critical decisions that have to be made by the winemaker.

European regulations allow the use of various forms of sugar at this stage. Chaptalisation has come to mean the use of beet sugar or cane sugar, both of which are composed of sucrose, the sugar used by M Chaptal. The other sugar, which is becoming more widely used, comes from the grape itself and is composed of concentrated, unfermented grape juice. This is sometimes used in its crude form, but more often after purification, when it is known as rectified concentrated grape must (RCGM). This is a colourless, syrupy liquid with no smell or flavour, composed almost entirely of a solution of glucose and fructose, the natural sugars of the grape. When this form of sugar is used, the process has come to be known as enrichment rather than chaptalisation. Officially, enrichment is either, and in practical terms there is no difference because sucrose is always converted to glucose and fructose by the acids in the must before being fermented to alcohol.

There is a considerable lobby to prohibit the use of sucrose for this purpose, the logical argument being that wine should be made entirely from the product of the grape. This would undoubtedly help to use some of the excess production of grapes throughout the world. However, the end result is identical, whether the sugars originate in grape, beet or cane: the debate is in the political arena.

One important fact should not be forgotten, and that is that enrichment has no effect on the sweetness of the finished wine. The sugar that has been added is fermented by the yeast to produce more alcohol.

Must concentration

As an alternative to increasing the sugar content of must by adding extra sugar, it is possible to achieve the same result by removing water. There are three main ways of doing this, all of which have to be communicated to the relevant wine authorities before commencing, and are strictly controlled according to the quality of the vintage – or so we are told!

The advantage that these procedures confer is that the really poor vintages of old no longer exist. In a bad year, it is possible to bring the concentration of sugars up to the required level for a reasonable wine, thus eliminating thin, acid wines.

The danger is that these processes become over-utilised, resulting in wines which are not characteristic and all tasting like old-style Australian Shiraz or California Zinfandel.

In all of the processes described below it is normal to concentrate a portion of the must and then add this back to the untreated bulk.

• *Vacuum distillation*

At atmospheric pressure, water boils at 100°C, a fact known by most people. Less well known is that if the pressure is reduced, water boils

Vacuum distillation unit at a producer of Barolo

at a lower temperature, and if the pressure is reduced as low as possible, water will boil at around 25 - 30°C.

If must were to be boiled at 100°C it would soon be reduced to toffee, a process which is actually used to make toffee. At 30°C the water can be boiled off with minimal effect on the remaining must. This is the oldest process used for the concentration of must, but it does have the disadvantage that some of the aromas are also boiled off. The newest generation of must boilers incorporate a chilled aroma trap that collects the aromas, condenses them and then adds them back to the concentrated must at the end of the operation.

• *Cryo-extraction (cryo-concentration)*

Another quite old technique used for the removal of water from grape must is to cool the must until it begins to freeze. The solid crystals consist of ice, which can be removed by filtration, leaving behind a concentrated must. This is a very simple method requiring only a cooling system, and there is virtually no loss of flavour compounds.

• *Reverse osmosis*

Osmosis is a natural process that is going on in our bodies at all times. All of our cells are surrounded by what is known as a semi-permeable membrane. This membrane allows the passage of water, but prevents the larger molecules from escaping. Nature tries to equalise everything, so if two solutions are separated by a semi-permeable membrane, the natural course of events is for water to pass across the membrane from the weaker to the stronger solution.

Reverse osmosis is the artificial situation that is produced by applying a high pressure on the stronger solution side, which forces water in the reverse direction, thus concentrating the strong side rather than diluting it. This is a good way of concentrating must as it is almost like filtering the water out under high pressure – and the aroma components are not lost either.

This technique is also used for the removal of alcohol for the production of low alcohol and de-alcoholised wines and beers, and also for the desalinisation of seawater. In the latter case, seawater is concentrated on the high-pressure side of a semi-permeable membrane, while drinking water emerges on the low-pressure side.

Reverse osmosis machine that can be used for concentrating and de-alcoholising

Nutrients

One of the essential elements for the healthy growth of yeast is nitrogen. This should not be confused with gaseous nitrogen (N_2), as used for blanketing wine to prevent the ingress of air to a vat. It is combined nitrogen in the form of ammonium (NH_4) compounds, which are necessary for the formation of amino acids and proteins within the yeast cell.

Most grapes contain an adequate natural level of such compounds, but if they are lacking there is a danger of the production of hydrogen sulphide, due to the powerful reductive nature of the fermentation process.

The simple remedy for the lack of nitrogen is the addition of an ammonium compound, such as diammonium phosphate (strictly, diammonium hydrogen phosphate) or ammonium sulphate, up to a total of 1g/litre.

Another nutrient additive is the vitamin thiamine, which is allowed at levels up to 0.6 mg/litre. Thiamine is an important vitamin in the growth of yeast populations.

Other treatments

• Bentonite

Occasionally bentonite is added at the must stage to remove some of the proteins in order to reduce the viscosity of the must, but care has to be taken that the level of nutrients is not reduced too far. Another advantage of bentonite at this stage is that it will partially remove the polyphenoloxidase enzyme, one of the powerful oxidising enzymes, thus protecting the must from deterioration by oxidation.

• Activated charcoal

If a must destined for the production of white wine is darkly coloured, it can be treated with activated charcoal up to 100 g/litre, to reduce the colour. But care is needed, as charcoal removes flavour as well as colour.

• Tannin

Grape tannins can be added to the must before fermentation, but care is needed to avoid a harsh taste in the finished wine.

Chapter 7

FERMENTATIONS

All love at first, like generous wine,
Ferments and frets until 'tis fine;

Samuel Butler 1692-1780

Fermentation processes are manifold, and if palatable wine is to be the result it is necessary to define this process accurately before considering the transformation of grape juice into wine. A biochemist would define a fermentation as any reaction involving either living microorganisms or, at least, an enzyme extracted from such an organism.

Antibiotics, vitamins, monosodium glutamate, citric acid and acetone are but a few products made on an industrial scale by fermentation processes. As wine makers and wine drinkers, we are interested in the alcoholic fermentation. This is a specific reaction during which yeasts feed on sugars and break them down with the aid of enzymes, producing alcohol and a considerable quantity of carbon dioxide gas and heat.

The accepted definition of wine is that it is 'the fermented juice of the freshly gathered grape'. This short phrase enshrines a surprising number of sensitive principles:

- Wine cannot be produced without an alcoholic fermentation. Even non-alcoholic wines are initially made by fermentation, but are then subjected to processes that remove the alcohol (and much of the flavour!).

- Wine is made only from grapes. The products made from other fruits must be referred to as 'fruit wine', and must bear the name of the fruit on the label.

- The grapes must be freshly gathered (although the definition of 'fresh' is given a somewhat liberal interpretation in the making of Amarone or Vin Santo). The origin of this clause goes back to the end of the nineteenth century, after the phylloxera disaster, when people were making 'wine' from imported raisins and concentrated grape must. It thereby excludes British wine, which is made from imported concentrated grape juice, but not English and Welsh wine, which is made from grapes grown in England or Wales.

Yeasts

Just as alcoholic fermentation is but one of a vast family of enzyme-related reactions, so the yeast used for alcoholic fermentation is a member of the huge family known as fungi, the same family that includes mushrooms and toadstools. Together with bacteria, they are responsible for the decay of all organic matter.

As a means of simplifying this complex array of microorganisms, it has sometimes been taught that yeasts can be divided into two groups: the wild yeasts and the wine yeasts. This, however, can lead to confusion. All yeasts are 'wild', in as much as they are all naturally occurring; they are indigenous. It is simply that some yeasts are better than others for the production of wine, so it is probably more constructive to think of them as 'strong' yeasts and 'weak' yeasts.

Yeasts are classified within fungi as *Ascomycetes*, or sac fungi, fungi whose bodily form is that of a small sac or capsule. Most wine yeasts belong to the genus *Saccharomyces*, meaning sugar-loving, of which the species *cerevisiae* is the commonest. Different strains of this yeast are used for wine making, brewing and baking. Other species are sometimes used in wine making, such as *Saccharomyces bayanus*, a yeast particularly associated with the flor in sherry production.

Yeasts from other genera also appear in fermenting musts, especially those that are being fermented by the natural flora. A commonly occurring member of this group is *Kloeckera apiculata*, a yeast with crescent-shaped cells, and intolerant of alcohol, so it quickly dies out as the fermentation progresses. Other species found in this group include *Saccharomycodes, Hansenula, Candida, Pichia* and *Torulopsis*.

Much discussion and argument has taken place about the role of yeasts in the production of different flavours in the finished wine. Some winemakers have maintained that the choice of yeast makes little difference. However, it is now generally accepted that yeast selection is very important, especially in wine for drinking young. Chemical analysis has shown that different yeasts will produce different amounts of various metabolites during fermentation. The list of these substances is considerable and includes simple compounds such as glycerol and acetic acid, and the more complex substances like carbonyl compounds, nitrogen and sulphur containing compounds, phenols, lactones and acetals. Considering also that many of these substances can interact, it is not surprising that the composition of the micro-flora has an important role to play in the flavour and complexity of the finished wine.

It is recognised that certain species are undesirable, such as *Brettanomyces* that can produce a farmyardy smell, so it does not seem unreasonable to assume that different species and different strains can produce different results, although these differences become smaller as the wine ages.

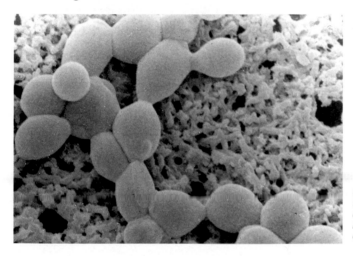

Cells of
Saccharomyces
cerevisiae
in process
of 'budding'

The various species of yeasts have differing tolerances towards some of the important components of wine. *S. cerevisiae* has the greatest tolerance towards ethanol, the predominant alcohol of wine, and has been known to produce up to 23% vol, whereas *Candid sp.* can manage only between 6 and 9%, and poor little *Hansenula sp.* die off when the alcohol reaches around 4%.

Likewise, the sulphur dioxide tolerance varies considerably between the species. It is one of those lucky quirks of nature that *S. cerevisiae* is one of the most tolerant. Thus a small dose of sulphur dioxide prior to fermentation gives this good yeast a head start over the other less robust organisms.

The action of yeasts

Yeasts are very adaptable organisms and can exist either aerobically, meaning in the presence of oxygen, or anaerobically, without it. In the aerobic state they reproduce rapidly, but do not produce any alcohol because they adapt their metabolism to extract all the energy from the sugars they are living on, producing water as an effluent. This situation can be simplistically represented as follows:

$$\text{yeasts} \ + \ \text{sugars} \ \rightarrow \ \text{water} \ + \ \text{carbon dioxide} \ + \ \text{heat}$$

This is approximately the same as any other organism living aerobically, such as human beings. We eat food, in its various forms, and excrete substances with very little nutritive value, and breathe out carbon dioxide from our lungs.

It is clear from this situation that yeasts actually prefer to be aerobic, because they are able to extract all of the energy from their foodstuff, reducing it down to the basic substances of water and carbon dioxide, which feed into the carbon cycle, and will ultimately be re-used by plants to produce more sugars.

The other indicator of aerobic preference is the speed at which the yeasts reproduce. They are obviously happy and reproduce at a high rate. This is the ideal condition at the start of a fermentation because the population increases rapidly, preventing any deterioration in the must. Therefore, aeration of the must before fermentation is one of the few occasions where oxygen is actually welcomed.

If the yeasts continued to have unlimited access to oxygen, no alcohol would be produced. But alcohol-free wine is not particularly attractive, as witness the several attempts at the creation of such products. So the yeasts must be encouraged to switch to their anaerobic mode, in which their metabolism is less efficient, and yields an alcoholic rather than an aqueous effluent. This change in behaviour occurs naturally because the dissolved oxygen in the must is quickly absorbed by the growing yeast population, resulting in anaerobic conditions.

The process of alcoholic fermentation can be summed up by the following:

$$\text{yeasts} \ + \ \text{sugars} \ \rightarrow \ \text{alcohol} \ + \ \text{carbon dioxide} \ + \ \text{heat}$$

It can be surmised that the yeast is not so happy in this condition, as its reproductive rate goes down, and it struggles with its metabolism. This process can be reduced to a proper chemical equation, which can yield useful information:

$$C_6H_{12}O_6 \rightarrow 2C_2H_5OH + 2CO_2$$

glucose   ethanol   carbon dioxide

180   92   88

In this equation the grape sugars are shown as glucose, but in reality they are roughly equal quantities of glucose and fructose. These substances have the same molecular formula, but the atoms in the molecules are joined together differently; glucose and fructose are said to be structural isomers (see p.22). The predominant alcohol in wine is properly called ethanol, although there are smaller quantities of many other alcohols, all of which play a part in the ultimate bouquet and flavour. In simple terms, this equation shows that each molecule of sugar produces two molecules of alcohol and two molecules of carbon dioxide.

Although this concept is useful when trying to understand the principles of fermentation, this equation is greatly simplified. Many other reactions are occurring, all of which add to the complexity of the finished wine.

The figures under each substance are the relative quantities involved in the reaction, by weight, and are obtained from the molecular weights of the substances. The units can be anything, provided they are the same on both sides of the equation. Thus, from this equation, it can be deduced that 180 kg of sugar will produce 92 kg of alcohol and 88 kg of carbon dioxide, or approximating, a given weight of sugar yields roughly half its weight as alcohol and the other half is lost as gaseous carbon dioxide (CO_2). This is a huge amount of gas, a gas that is colourless, heavier than air, and is highly dangerous because it can fill empty vats and will cause immediate suffocation if inhaled in high concentration. It is important that a good ventilation system is operational during fermentation, although in many older wineries this amounts to opening all the windows as wide as possible!

The figures in the above equation are all by weight, but alcoholic strength is always expressed by volume. To convert alcoholic units from weight to volume it is necessary to involve the density of alcohol in the calculations. Ethanol is less dense than water, weighing only 0.7897 grams per millilitre (water being approximately 1 g/ml). Using this fact, it can be calculated by simple mathematics that 17 grams of sugar per litre of must produces one percent of alcohol by volume. By using this formula in conjunction with the results of the sugar analysis, the winemaker can calculate how much sugar to add at the enrichment stage.

For example, if a winemaker wants to make a wine at 12% vol alcohol and the analysis of the must shows it to contain 170 g/l of sugars. How much sugar must be added at the enrichment stage?

• It is known that 17 g of sugar per litre yields 1% vol of alcohol

• Therefore 170 g/litre will yield 10%vol

• The increase in alcohol required is 12 − 10 = 2% vol

• Therefore, quantity of sugar to be added is 2 x 17 = 34 g/litre

• For 4000 litres of must, total quantity of sugar
                                to be added is 4000 x 34 = 136kg

However, as stated above, it should not be forgotten that this calculation is based on a simplified view of the fermentation reaction, and this figure is only approximate.

Natural fermentation

In natural fermentation (variously known as spontaneous or traditional fermentation), use is made of the natural micro-flora that are found on the grapes, in the vineyard and on all the surfaces of the winery. This mixture of organisms consists of all the various yeasts and bacteria that are indigenous. In those parts of the world where wine has been made for centuries, the yeasts have gone through a process of mutation and natural selection, resulting in an adequate supply of good wine yeasts in the micro-floral blend.

At the beginning of the fermentation process most of the bacteria are knocked out by the sulphur dioxide which will have been added, and the most abundant yeast will start the fermentation. Some of these yeasts start to die at an alcohol concentration of around three per cent because they are intolerant of alcohol. The fermentation is completed by the more powerful yeasts, probably varieties of *Saccharomyces cerevisiae,* that will vary from vineyard to vineyard and even from year to year. It is this unpredictable and somewhat erratic nature of natural fermentation that has resulted in the reduction of the number of wines produced by this ancient process.

However, there is a renewed interest in this subject, as it has been realised that the great and valuable attribute of natural fermentation is that it imparts character and individuality to the wine, due to the complex mixture of organisms that are present. These are sometimes referred to as 'wild ferments' because they rely entirely on the natural range of microorganisms found in the wild.

Cultured yeasts

Studies in wine research have shown that the nuances imparted by the yeast are quite considerable, although this effect is most marked in the earlier stages of maturity. Hence the interest in cultured yeasts, which offer the winemaker the choice of the characteristics of a particular yeast. The quality-conscious winemaker selects yeast carefully, often after many years of experimentation.

Cultured, or selected, yeasts have been grown from samples that have been taken from vineyards around the world, where wines have been produced for centuries and where there is a reasonable chance of finding top quality strains. It is a simple task for a microbiologist to separate individual colonies from a culture of mixed organisms and to grow each colony into a working population. This can be purchased either in a freeze-dried form or as an actively living culture on small slopes of agar, a nutritious jelly that will support live yeasts. The advantage of the freeze-dried form is that it has a longer shelf life and can easily be re-activated by mixing with fresh must.

Using a cultured yeast is simple, and does not necessitate the sterilisation of the entire vat of must because the success of one yeast over another depends on population density. The yeast with the greatest population wins (although it is possible to go still further and use what is called a 'killer' yeast, which produces toxins that destroy other yeast strains). Having purchased the culture, all that is necessary is to inoculate it into, say, 100 ml of must sterilised by filtration, put it in a warm place and allow it to grow until actively fermenting. This is then poured into a larger volume of must and left to develop. This in turn is added to the vat of juice, stirred and left to ferment. The actively developing added yeast overwhelms any other yeast that might be present in the must.

The great joy of the wines of the world is their infinite variety, and one of the great drawbacks of this technique is a standardisation of style: the production of 'industrial wines' that lack individuality, always the same, always perfect and with no indications of origin. There can be no doubt that wines have moved closer together in style since the advent of modern winemaking.

The ultimate technique employed by the more ingenious winemakers is to grow a culture of microorganisms collected from their own vineyard, by employing the services of a microbiologist to separate and culture each individual yeast. Trial fermentations are then carried out with the separate yeasts to discover which produces the best results. Thus, we get that valued combination of individuality with reliability of production.

Control of temperature

Yeast is a living organism and, like us, produces heat when active and slows down when cold. A great deal of heat is produced during fermentation, which has to find an escape route. It is amazing to feel the heat radiating from a vat whose contents are in active fermentation – it is not surprising that cellar workers gather in this area during their tea breaks! Conversely, if the fermenting liquor becomes too cold, as can happen in chilly autumns in northern Europe, the fermentation will become too slow, or even come to standstill.

Temperature is an important factor during fermentation because two objectives which depend upon opposing conditions come into play. These are the extraction of flavour and colour components from the skins on the one hand, and the retention of volatile aromatic substances on the other. Warmth is required for the former, whereas for the latter, the cooler the better.

For each wine there is an optimum temperature, balancing skin extraction with aroma. In the production of red wines, both of these objectives are important, and the choice of temperature is critical. Skin extraction, however, plays no part in the fermentation of white wines, because the skins are not present during the fermentation stage, the only consideration here being the retention of volatile aromas. Even if skin contact techniques have been employed, the skins are removed before fermentation commences (see p.104). This is the basic reason that white wines are, in general, fermented at lower temperatures than red wines.

The control of temperature occurs naturally in a traditional cellar because the small size of the vats and barrels gives a high surface to volume ratio, and thus an efficient mechanism for heat loss. Additionally, in northern Europe, the end of summer comes quickly, autumn is cold and the cellars become even colder, so there is usually little that needs to be done to remove the heat of fermentation.

When larger vats are used, or where the cellars are warm and dry, artificial cooling is necessary because the surface area of large vats is insufficient to dissipate the large amount of heat produced. Vats such as these have cooling coils inside and jacketed walls, through which is circulated a refrigerant, automatically controlled by thermostats.

The usual range for the fermentation of red wines is 20°C to 32°C, whereas whites are normally fermented at temperatures of between 10°C and 18°C. But, as with all winemaking activities, rules are frequently broken. Some red wines are fermented above 32°C and at least one brilliant producer of Muscat de Beaumes de Venise has persuaded his yeast to ferment at 0°C, despite all that has been written about yeasts not working below 5°C!

In recent years there has been a fashion for cold fermentation of white wines because it has been realised that this technique gives an improvement on the dull beverages that used to be offered by the old-fashioned wineries, particularly in warmer climes. The result has been a proliferation of pale green wines tasting as if they were flavoured with essence of pineapple and banana. This flavour is produced by the retention of esters that would normally be lost at higher temperatures and which impart a sameness to all of these products of low temperature fermentation. The ideal is to find the sensible balance between the warm conditions that result in the loss of aromatics and the excessively low temperatures that produce the tropical fruit versions. It was interesting to note that a famous producer of high quality wines in Alsace ferments all of his white wines at 18°C to maximise the varietal aromas.

Monitoring the fermentation

The rate of fermentation and the temperature of the fermenting liquid are inextricably linked: the higher the temperature, the faster the fermentation. The only means for controlling the rate of fermentation is by the control of temperature. In traditional small scale wineries, nothing need be done other than being prepared to take emergency measures should the ambient temperature become abnormally high or low.

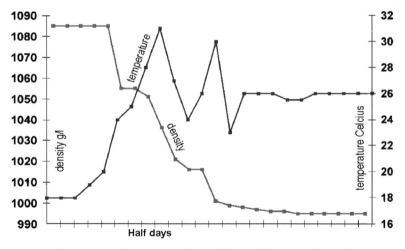

Graph showing the progress of fermentation by following density readings

In a modern winery, life should be even easier because temperature control can be automatic. With state-of-the-art equipment, it is easy to hold the temperature of even the largest vats to a remarkably constant degree. This, of course, requires the installation of high capacity cooling equipment. It is the lack of such capacity that often prevents the making of the highest quality wine in hot climates.

The progress of fermentation can be monitored simply by measuring the density of the fermenting liquid, as in the graph above. Unfermented must is more dense than water, and alcohol is much less dense, so the transformation from must to wine is accompanied by a considerable drop in the density. By measuring the density regularly, it is possible to construct a graph of the progress of the fermentation. Adjustment of the temperature can then be used to control the rate of fermentation. The graph above shows the typical fluctuations that take place in a simple winery with basic cooling equipment. Years of experience have shown that, provided the temperature is maintained within a given range, these fluctuations are of no consequence.

Another way of monitoring the situation is to check the population density of the yeast cells by doing a cell count. This requires a haemocytometer or counting cell and a microscope, which are somewhat more expensive than a hydrometer. This is the really scientific way of monitoring a fermentation, as the population density is of prime importance. The rate of change in density of the fermenting liquid is proportional to the population density.

Stopping the fermentation

Fermentation comes naturally to a halt when most of the sugars have been exhausted, although even the driest of dry wines will still contain a tiny quantity of sugar amounting to around 2 g/l. This natural termination of fermentation is the route followed by all winemakers of the old tradition – and the wine is essentially stable. It contains no fermentable sugars, it is depleted in yeast nutrients and it will not re-ferment. Therefore it does not require modern bottling techniques and only needs protection from the atmosphere to prevent oxidation and the development of volatile acidity.

Although many fermentations are allowed to come to a natural conclusion, commonly known as fermentation to dryness, this takes a long time, longer than commercial wineries can afford. The intervention of science has provided several ways of terminating fermentation prematurely.

1. Increasing the pressure of the carbon dioxide

Much carbon dioxide is produced during fermentation. If this is trapped above the wine by fermenting in a pressure vessel with the outlet valve closed, the pressure will rise. At about seven atmospheres the yeast will cease activity because it is effectively being suffocated by its own carbon dioxide. If the pressure is allowed to drop, the yeast will start the fermentation again. Therefore, increasing the pressure is only a temporary way of halting fermentation.

2. Reducing the temperature

In common with most other forms of life, yeast activity becomes slower as the temperature drops, until it ceases altogether at around 5°C – unless a super-cool yeast is present (see above). However, this is only a temporary measure, and fermentation will re-commence if the temperature rises. This is the method used to stop the fermentation during the production of Asti (formerly known as Asti Spumante).

3. Killing the yeasts

In earlier times, when sulphur dioxide was splashed around in a carefree manner, some of the cheap desert wines were made by adding excessive amounts of this noxious chemical to kill the yeasts and arrest fermentation. With the stricter control of additives, this is no longer possible because the maximum sulphur dioxide allowed is much lower. A more acceptable way of killing yeasts is pasteurisation, the heat treatment widely used for killing microorganisms in milk. In this process, the liquid being pasteurised is heated to around 80°C for a few seconds, which destroys all the yeasts.

4. Removing the yeasts

A simple way of stopping fermentation is to remove the yeasts completely by filtration, or by using a centrifuge, which physically separates the solids from the liquid by centrifugal force. Once again, this is only a temporary effect; if a few yeasts get back into the wine the fermentation will start again.

5. Adding alcohol, or fortification

Spirit can be added at an appropriate point during the fermentation, as in the production of port and *vins doux naturels*, to raise the alcohol level above that which the yeast can tolerate, and it dies.

Of all these methods of stopping fermentation, the only one that results in a stable wine is fortification. The product of all the other processes is unstable in that fermentation can be re-established by the introduction of more yeast. To prevent re-fermentation in bottle, these wines have to be bottled by modern techniques, which means without any troublesome microorganisms.

A 'stuck' fermentation

The situation that all winemakers dread during the production of a normal dry wine is when the fermentation stops unexpectedly, before all the fermentable sugars have been consumed. This can happen if there is a sudden drop in temperature, or if there is a lack of nutrients for the yeast, such as vitamins or amino acids. The problem in this case is persuading the yeast to come back to life, for it is frequently reluctant to do so. Merely warming the vat or barrel often has little effect.

A cascade for oxygenating fermenting must

Sometimes pumping the must over a cascade to dissolve some oxygen does the trick, by giving the yeast a quick lift into its aerobic mode. Perhaps the addition of some yeast nutrient will bring them back to life. The only sure way of completing the fermentation is to gradually blend the 'stuck' must into another vat of must which is in active fermentation.

Naturally sweet wines

Yeasts will normally carry on fermenting a must until all the sugars have been consumed, so it might be wondered how the great sweet wines of the world, such as Sauternes or Tokaji Aszú, can be produced. What actually happens is that the rate of fermentation becomes extremely slow by virtue of several different factors, and is finally stopped by cooling the wine and by the addition of sulphur dioxide.

First, there is a progressive weakening of the yeast due to the very high concentrations of sugar combined with the increasing level of alcohol. This results in the shrinking of the yeast cells by osmotic pressure. This is the principle by which nature tries to even things out. If two solutions are separated by a semi-permeable membrane, such as surrounds every cell in our bodies, water tries to move across the membrane from the weaker solution to the one of higher strength, thus making them equal. Therefore, if a yeast finds itself in a strong solution, its cells become dehydrated as its own water is sucked out.

Second, the nutrients are gradually exhausted as fermentation proceeds, making it more and more difficult for the yeasts to survive.

Finally, if the original grapes were affected by *Botrytis cinerea*, otherwise known as 'noble rot' or *pourriture noble*, the must will contain substances that have anti-fungal properties and which will inhibit alcoholic fermentation.

The malo-lactic fermentation

When the yeast fermentation, otherwise known as the alcoholic or primary fermentation, has come to an end, it is not the end of fermentations because another one frequently takes place that involves a different set of microorganisms.

Following on from the primary fermentation is the malo-lactic fermentation (MLF), also known as the secondary fermentation. (Note that this term has often been misused in relation to the faulty condition of wine fermenting in the bottle. This latter condition should be known as a second fermentation, or a re-fermentation, not secondary fermentation.)

The secondary fermentation used to be regarded as one of the great mysteries of winemaking. It had long been recognised that the character of the wine changed during the period following the end of the primary fermentation, becoming smoother and generally more pleasant. Unfortunately, because the mechanism was unknown, it was not possible to control the result. Although Pasteur discovered the action of micro-organisms in the nineteenth century, the mystery remained unsolved until the middle of the twentieth century, when Ribéreau-Gayon in Bordeaux and Ferré in Burgundy both discovered that the malo-lactic transformation was a fermentation involving a mixed culture of lactic acid bacteria.

We now know that the malo-lactic fermentation is the result of bacteria attacking the harsh-tasting malic acid and transforming it into the softer-tasting lactic acid. Once the mechanism had been established it became possible to control the result at will. The basic reaction is:

$$HOOC.CHOH.CH_2.COOH \rightarrow HOOC.CHOH.CH_3 + CO_2$$

malic acid   lactic acid   carbon dioxide

The carbon dioxide produced during the reaction disappears into the atmosphere unless the wine has already been bottled, in which case it renders the wine *pétillant*. This was the original source of the sparkle in vinhos verdes, although nowadays there are other, cheaper, ways of achieving the same result (see p.117, pompe bicyclette!).

The softening of the acidity is advantageous in red wines, giving a smoothness and a suppleness that might have been absent prior to the action of the bacteria. Most winemakers ensure that all of their red wines have completed the malo-lactic fermentation before bottling, thus preventing the unpleasant taste of the active fermentation and the prickle of the dissolved gas developing in the bottle.

White wines that are intended to be full-bodied and soft, such as many of those made from the chardonnay grape, are often encouraged to undergo the malo-lactic fermentation (provided they have sufficient initial acidity to prevent them from becoming 'flabby'). Those wines that are naturally crisp, such as products of sauvignon blanc, are usually preferred in their natural state. Sometimes in the case of a very acidic wine it is useful to enlist the aid of the secondary fermentation to reduce the acidity to an acceptable level. The paradox here is that in certain wines with very low pH, where the reduction in acidity would be most welcomed, the bacteria are reluctant to grow as they are pH sensitive. In this case it is necessary to carry out a chemical deacidification in order to decrease the acidity to a level at which the organisms will grow.

The bacteria involved are quite varied but all produce the same result: the destruction of malic acid and the production of lactic acid. They include various species and strains of *Lactobacillus, Leuconostoc* and *Pediococcus,* many of which are naturally present in the fermenting wine and will start spontaneously. However, if necessary, these can be purchased as cultures and are simply grown in the winery and added to the wine, either during the primary fermentation or after it has finished. In this case, the organism usually involved is *Oenococcus oeni.* Being bacteria, they grow faster at higher temperatures and are readily killed by sulphur dioxide. Thus, the control of the secondary fermentation is simple: if it is wanted, the bacterial culture is added, the temperature of the wine is raised and sulphur dioxide is not added. Conversely, the prevention of the malo-lactic can be achieved by adding sulphur dioxide and keeping the wine cold.

The results of the MLF are generally looked upon with favour because not only does it reduce acidity, but it also increases complexity by generating additional flavour components such as diacetyl, the source of the buttery character. It also increases the stability of the wine by consuming bacterial nutrients. This is particularly useful in red wines as they lack the protection of added sulphur dioxide, which becomes bound to the anthocyanins, the colouring pigments (see above).

However, there is sometimes a loss of fruit owing to a reduction in some of the fruity esters, which are broken down by the action of the bacteria. In addition, the buttery flavour can be too dominant in some wines. It is fortuitous that control of the MLF is possible, although always with the risk that the fermentation will start at some time in the future, after the wine has been bottled. However, this is unlikely with modern bottling techniques (see p.217).

Chapter 8

RED & PINK WINE PRODUCTION

Great fury, like great whisky, requires long fermentation.

Truman Capote 1924-84

Alcoholic fermentation is but one part of the totality of operations that constitutes the making of wine. It is the expertise of the winemaker that governs the manner in which wine is made, and the choice of the processes used. These will vary according to the style of the wine being produced. At every stage the winemaker can choose from a vast array of different techniques, and it is this choice that makes the difference between wine and great wine.

The sequence of events in the making of red wine involves some or all of the following operations, in the order in which they are carried out:

- de-stemming
- crushing
- sulphiting
- adjustments to acid and sugar
- heat treatment
- maceration on the skins
- fermentation with the skins
- post-ferment maceration
- draining
- pressing
- clarification

The principal aim in the production of red wine is the extraction of colour and flavour from the skins, as the juice of virtually all black grapes is colourless. This extraction is normally achieved by a combination of maceration and fermentation, although sometimes heat is also used to weaken the cell structure and release the polyphenols more readily. There is one notable exception in the Alicante Bouschet, which has deeply coloured red juice and produces a black wine which can be used to deepen the colour of anaemic wines.

When grapes are crushed, the yeast on the skins mixes with the juice and starts a fermentation, producing alcohol, which in turn extracts the polyphenols from the skins. This is a very effective and simple way of releasing the colour and flavour, but the major problem with this process is the production of a large volume of carbon dioxide, which is evolved as a myriad of small bubbles. These bubbles become entrapped in the solid matter, floating it to the surface.

At this point in the winemaking process the opposite is required, and trouble will quickly occur unless steps are taken to submerge this floating cap because the skins are swarming with acetic bacteria. In the warm and damp surroundings of the fermentation cellar this will soon turn the wine to vinegar. Furthermore, a floating cap of skins will not be in intimate contact with the liquid and will not be efficiently extracted by the liquid, so will not yield its full measure of flavour and colour.

Fermentation vessels

The size of the container used for fermenting grape juice has steadily increased as wine has become more popular and wineries have grown in size.

Fermentation of red wine usually takes place in a vat, which originally would have been made of wood and of the order of 50 hl in capacity. At this size temperature control would have been automatic, as the ratio of surface area to volume would have been correct for the dissipation of the heat of fermentation, especially in the cold cellars of northern Europe. Oak vats are still highly regarded because they give good results.

As volumes increased it was necessary to construct vats of other materials, the next in evolution being those made of cement. These are still widely in use today, and there is nothing wrong with them provided they are lined with tiles or epoxy resin. Neither of these materials is perfect: tiles come away from the walls, creating nasty 'bug' traps that cannot be sterilised, and epoxy resin needs renewing from time to time.

But such is the love of these vessels, because of their thermal stability, that some winemakers have cut slits in the side walls to enable sheets of stainless steel to be inserted and then welded together inside the vat to create a vat in a vat!

Modernised cement vat with stainless steel lining.
Note the temperature control panels and the tartrate deposit.

Modern wineries are equipped with stainless steel vats, but even these are not perfect. The thin steel walls confer no thermal stability and temperatures can fluctuate wildly. However, they are available in all sizes and shapes, can be easily cleaned, and come already fitted with temperature control panels.

Temperature control is the main problem with large vats because the considerable amount of heat that is generated during fermentation must be carried away to prevent the temperature getting out of control.

It is now possible to purchase tanks specially designed for the production of red wine. One example of this is the Ganimede tank where the pressure of the carbon dioxide pushes some of the fermenting must into a separate compartment at the top of the tank. When a pressure release valve opens, this wine floods back into the main part of the tank and submerses the floating skins.

Maceration

One of the principles of the production of red wine is understanding how to handle the processes involved in the extraction of colour and flavour from the skins of the black grapes. The all-important fact is that the skins have to be handled with care in order to extract only those elements that are required and to avoid those that result in a bitter, tannic wine, totally lacking in charm. The more the skins are agitated, the greater will be the extraction of the unpleasant fraction. This requires skill and experience.

Pre-ferment maceration, otherwise known as a 'cold soak', can be used to extract more aromas from the skins. This is identical to the so-called 'skin contact' process as used in the production of aromatic white wine (see p.104). During this period the must has to be cooled to somewhere between 15 and 4°C in order to prevent the fermentation starting, so that the cells containing the flavour and aroma compounds can be broken. This is particularly effective with Pinot Noir, where the aromas are very valuable but the danger of the extraction of polyphenols is minimal because of the nature of the thin skins.

Post-fermentation maceration is used to extract more polyphenols for enhancing the ageing characteristics of a wine, but this has to be done carefully if the wine is not to become overly tannic and bitter. As opposed to the pre-soak, this maceration is carried out at the relatively high temperature that is naturally achieved at the end of the fermentation. (This is where the old-fashioned wooden or cement vats are preferred because their thermal insulation has the advantage over the more modern stainless steel tanks.) During this period, apart from the extraction of more polyphenols, the tannins and anthocyanins interact, producing more stable colours that add to the ageing potential of the wine.

Traditional process

The earliest and the simplest method is to pump the mass of crushed grapes and juice into an open vat and to allow it to ferment naturally. When the evolution of carbon dioxide begins, the gas bubbles become trapped in the solids, which are raised to the surface. Here they form a compacted layer and eventually rise out of the liquid, forming a cap, known appropriately by the French word *chapeau*. This is a dangerous situation, with the combination of acetic bacteria, warmth and oxygen poised to convert the vat to vinegar overnight (see p.4). So the cap has to be punched down manually several times each day (Fr. *pigeage)*, which is a tedious process and is really only suitable for small artisan wineries and for small volume fermentations of short duration. Although tedious, it is actually a gentle process causing minimal disturbance to the skins and avoiding the harmful abrasion which would release the hard tannins from the outer layer of the skins cells.

A modern mechanical and automatic machine for punching down at Cave de Tain

The technique involves balancing on the edge of the vat, holding an unwieldy pole, on the end of which is attached a disk with which the floating layer of skins is punched down beneath the surface. In these days of health and safety awareness it is somewhat surprising that this operation is still permitted because people have fallen into vats of fermenting wine whilst punching down the cap, several with fatal results. One might be tempted to comment "What a way to go!", but the loss of life is obviously not an occasion for joking, and in any case, wine at this stage of its production is not a particularly pleasant beverage.

Such is the regard for punching down that some modern wineries have installed automatic mechanical equipment for carrying out this process. A stainless steel cone attached to a hydraulic piston is positioned above the vat and gently submerges the skins in all directions. Then, after about twenty minutes, it moves on automatically to the next vat, and this is repeated all day long. Another version of this machine uses the principle of a propeller that rotates and causes a powerful current that submerges the skins. Tiredness does not set in, neither does anyone get drowned.

Submerged cap process

One approach to the problem of rising skins is to keep them submerged by placing a perforated screen just under the surface of the liquid. This prevents the skins from being pushed out of the liquid and eliminates the danger of the development of volatile acidity (see p.238), but it is an inefficient process because the skins become compressed under the screen, preventing efficient extraction. The relentless production of carbon dioxide compresses the skins ever more tightly against the screen,

A simple arrangement of perforated screens to hold down the skins of Pinot Noir during fermentation. The massive wooden structure indicates the large forces involved.

compacting them and preventing the gas escaping, until eventually something has to give. Indeed, vats have been known to burst, resulting in the total loss of the contents.

An ingenious vertical stainless steel vat has been designed that overcomes this problem by incorporating a perforated riser pipe to permit the escape of gas from the body of the ferment. Being made of stainless steel, this design also incorporates automatic temperature control by means of cold brine running down the outside of the vat.

Pumping-over systems

An alternative approach to the mixing of skins and liquid is to pump the liquid from the bottom of the vat over the skins floating at the top. An effective extraction takes place as the liquid percolates through the skins. In its simplest form this can be achieved in an old-fashioned concrete vat by pumping the juice from the bottom of the vat through a flexible hose that can be manually directed over the floating skins.

Manual pumping-over in the production of Aglianico del Vulture

In a more modern stainless steel installation, this can be achieved automatically by means of a fixed spray-head in the top of the vat, which is fed by a pump taking liquid from the bottom. The wine is protected from oxidation by the constant evolution of carbon dioxide from the fermentation.

Château Belair in Saint-Emilion uses a system which is the epitome of gentleness. Juice from the bottom of the fermenters is run by gravity into a small tank which is then raised by an electric winch so that the juice can flow back again by gravity over the skins.

Large systems using the pumping-over principle have been developed with specially designed vats, and are often used for making wines of medium to high quality. These vats are double the height of an ordinary vat, with a perforated screen that divides the vat into two parts. The crushed grapes are introduced into the upper part, from whence the juice drains into the lower section, where it ferments. A pump transfers the juice from the bottom section of the vat to the top of the vat, where it sprays over the skins and gently percolates down to the bottom section again, extracting the colour and flavour en route without disturbing the mass of skins. There is no danger of dissolving atmospheric oxygen as this takes place in a closed vat.

Délestage (Rack and return)

The *délestage* process could be regarded as an extreme version of pumping-over, whereby once per day the entire volume of liquid in the fermenting vat is drained off into a second vat and is then returned to the original vat by spraying it over the skins.

This procedure achieves several different goals:

*   The fermenting must is well aerated, thus stimulating the metabolism of the yeast.
*   The anthocyanins and tannins combine to form stable colouring compounds, and the lower molecular weight tannins polymerise to form tannins with a softer taste.
*   The heat generated in the cap of skins is re-distributed.
*   Pips can be removed by incorporating a suitable screen, thus minimising the extraction of bitter tannins.

The result of this process is a wine that is smoother and more drinkable within a few weeks of the harvest.

Autovinifier

The Algerian Ducellier system, or autovinifier, is really another form of pumping-over, but is ingenious in that it needs no external power because it harnesses the energy from the fermentation in the form of the carbon dioxide pressure to move the juice to the top of the vat. It was for this reason that it was developed in Algeria, where there was no electrical power in the distant areas of that vast country. The idea was transplanted to the Upper Douro for the same reason, although it is ironic that the Douro is now the source of vast amounts of electrical power from its hydroelectric schemes. Such is the efficiency and simplicity of the autovinifier that it can be found even in modern winemaking countries such as Australia.

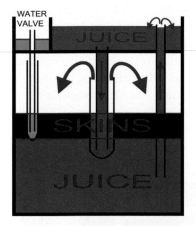

The autovinifier consists of a concrete vat in the form of a cube with an open concrete trough on the top, fitted with a tube through which juice rises into the trough and a siphon through which it falls back into the main chamber. The crushed grapes are loaded into the chamber, which is then sealed. When the fermentation commences, the pressure of the trapped carbon dioxide causes the juice to rise through a tube into the trough at the top. When the level of liquid in the bottom chamber falls below the level of a gas escape tube, the pressure drops and the liquid in the trough cascades back into the lower chamber. As it does so, it thoroughly churns the skins and extracts the polyphenols in a very efficient, if somewhat rough, manner.

The original Ducellier tanks are still widely used both in Algeria for the production of good quality light wine, and in the Upper Douro where they are used for both port and light wine production.

Autovinifiers dismantled for cleaning. The syphon tubes and water valves are lying on top of the tanks.

Rotary fermenters

The rotary cement mixer has long been known for its simplistic and effective way of mixing solids and liquids. It was a short step to develop this principle for the mixing of grape skins and must. The blades inside the tank are designed in such a way that rotation in one direction results in a mixing of the contents; rotation in the opposite direction empties the drum. Its action is very effective because it achieves a thorough mixing of skins and juice and has found favour in the vinification of difficult grapes, such as pinot noir, whose thin skins contain a meagre level of colouring matter. However, this severe disturbance of the skins can lead to over-extraction, so expertise is vital in its operation.

Rotary fermenters used for Pinot Noir wines

Thermo-vinification

It is possible to extract the necessary components from skins by the use of heat, rather than the effect of fermentation. The method, as practised in eastern European countries in particular, is to heat the crushed mixture to 60 - 75°C, holding it at this temperature for 20 to 30 minutes, and then to cool it down to fermentation temperature. This gives an intensely coloured must because the anthocyanins in the skins are readily extracted at elevated temperatures, owing to the weakening of the cell walls by the heat. This can result in a rather cooked flavour in the wine unless care is taken to use only the minimum amount of heat. Another disadvantage is that the colour is somewhat unstable and is partially lost during the fermentation

A good example of an ingenious adaptation of this principle is used at Château de Beaucastel, where all of their black grapes are subjected to a heating process. After de-stemming, the whole grapes are passed rapidly through a heat exchanger at 90°C, which heats only the surface layers of the grapes, the juice remaining cold. This weakens the skin cells without damaging the delicate flavour components, so that in the subsequent maceration the anthocyanins are easily extracted. At the same time it destroys the polyphenoloxidase enzyme that causes premature oxidation, thus requiring a lower dose of sulphur dioxide. (For more information see www.beaucastel.com)

Flash détente

A process that was first introduced in 1993 for the production of aromas from bananas, mangoes and lychees has been adapted for the production of wines rich in polyphenols and aromas. The grapes must be well-ripened and stripped of all stems to avoid unpleasant flavours. They are heated to a maximum of 95°C for several minutes in an oxygen-free atmosphere and are then immediately subjected to a low vacuum where the cells are ripped apart. The polyphenols and flavours are readily extracted during the maceration that follows, producing a wine so rich that it has to be blended with that from a more traditional process.

It has been authorised by INAO for AOC production and is becoming widely used in the south of France for Côtes du Rhône wines and wines from the south-west of France.

Carbonic maceration (Macération carbonique)

Carbonic maceration is a widely used process, producing a wine that is ready for drinking in a matter of weeks rather than months or years. This makes it the accountants' dream as it can have a very positive effect on cash flow. Financial considerations aside, it produces a style of wine that is popular, a wine that is of good purple colour, fresh and fruity on the nose, soft and eminently quaffable, and ready to drink.

The two important factors necessary to the success of this process are the use of whole bunches of undamaged grapes and a fermentation vessel that can be filled with carbon dioxide. The old name for carbon dioxide is carbonic acid gas, hence the name of carbonic maceration.

Any vat can be used for carbonic maceration, provided it can be closed at the top to prevent ingress of air. The fermentation takes place in two distinct stages, the first in a specially prepared closed vat at an elevated temperature, the second in an ordinary vat at normal temperature. The main principle of carbonic maceration is that the first stage, the true carbonic maceration, takes place without the involvement of yeast. The alcohol that is produced during this phase is formed by intracellular fermentation inside the grape, using the grape's own enzymes, and it is for this reason that the grapes must be whole.

Whole bunches of undamaged Grenache for making Coteaux de Languedoc at Cases de Pène

During the second phase, the more usual extracellular fermentation takes place, using the yeasts from the skins of the grapes. In the past this process has been misunderstood, so it is worth noting with care the main principles at each stage.

The complete sequence of events is:

1. A vat that can be totally closed (usually a concrete vat with metal hatch covers) is flushed with carbon dioxide to sweep out all the air, thus eliminating all of the oxygen. It should be noted that the vat remains at atmospheric pressure throughout the process, but air cannot enter.

2. The vat is filled with whole bunches of undamaged grapes. It is important that the grape skin is intact, otherwise the juice will escape and an ordinary fermentation will commence.

3. The lack of oxygen and the presence of carbon dioxide cause complex metabolic changes to occur inside the grape, which can be simplistically associated with dying. The structure of the grape is attacked by the grape's own enzymes, releasing the sugars from within the cells. The grapes are undergoing anaerobiosis.

4. The enzymes in the grape attack the sugars, breaking them down to alcohol. This is **intracellular** fermentation, and takes place in the absence of yeast (although it is the same biochemical process that occurs inside the yeast during normal fermentation).

5. The biochemical reactions cause the temperature to rise to between 30 and 35°C. This process is allowed to continue for between five and fifteen days, during which time about 3% alcohol is produced.

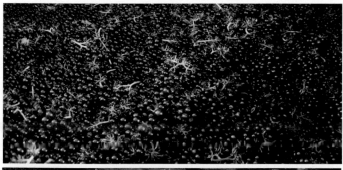

Freshly picked Gamay grapes

After one week they have liquefied and become Beaujolais Nouveau

6. At the end of this period, the contents of the vat become soft and semi-liquid, and the characteristic aromatic flavour will have developed. The vat is drained, the juice separated, the skins pressed as usual and the press juice added to the free-run juice. Another

unusual characteristic of carbonic maceration is that the press juice is of better quality than the free-run juice.

7. The combined juice is cooled to about 20°C, transferred to an ordinary vat and the fermentation allowed to run to completion, using either the natural yeast from the grape skins or an added selected yeast. This is **extracellular** fermentation.

The result is a combination of the best of both worlds: good extraction of softened skins aided by the elevated temperature, followed by the completion of fermentation at a lower temperature, which retains the volatile aromas.

Variants on carbonic maceration

True carbonic maceration is as described above, and is used in many countries around the world for the production of light and fruity wines, ready to drink after only a few months. These wines are low in tannins and polyphenols and have a comparatively short life. However, there are other procedures that make use of intracellular fermentation as part of the fermentation process.

• Whole bunch fermentation

Whole bunch fermentation is typified by the production of Beaujolais, where whole bunches of Gamay grapes are tipped into open-top vats that have not been previously flushed with carbon dioxide. Those at the bottom of the vat become crushed and the juice oozes out, coming into contact with the yeast on the skins, starting an ordinary extracellular fermentation. This produces a large volume of carbon dioxide, which, being heavier than air, gradually pushes the air out of the top of the vat and swathes the whole grapes in carbon dioxide, which promotes intracellular fermentation. These wines are thus the result of a combination of extra- and intra-cellular fermentation of somewhat unknown proportions. It is the combination of these fermentation conditions with the fruit of the Gamay grape that gives Beaujolais its unique characteristics. This process is known as *semi-macération carbonique*, or as a Beaujolais producer called it: *"Macération Beaujolaise"*!

The different styles of Beaujolais are dependent upon the length of time that the berries are left to macerate. Beaujolais Nouveau is racked off after about one week, whereas the higher grades are left for a longer period of anything up to a month.

• *Whole berry fermentation*

Another variant is the use of whole berry fermentation. In this process the bunches are de-stemmed but the berries are not crushed. Intra-cellular fermentation takes place inside the whole berries, while the normal yeast fermentation continues in the juice that has escaped during the filling of the vat. The resultant wine has a freshness imparted by the intra-cellular fermentation, combined with the structure obtained from the maceration of skins in a fermenting liquid.

Pink wines

It is unfortunate that the French term rosé has been applied to all pink wines from any country. The correct term for Spanish pink wines is *rosado* or *clarete* and for Italian *rosato* or *chiaretto*, or *cerasuolo* in Abbruzzo. We don't call red wines 'rouge' or whites 'blanc', so why should English speaking nations not use the term 'pink'?

Pink wines have had a bad reputation due to the over-sugared old style Rosé d'Anjou, Portuguese Rosé and Blush wines from California. Modern pink wines are deliciously fresh and dry, and redolent of summer fruits.

To dispel a widely held belief, it should be noted that most pink wines are not made by adding red wine to white. The only wine made by this method is pink champagne; in Europe it is forbidden to make pink wine by blending red and white wines. This is a very apt piece of legislation because the flavour of a wine made by extraction of the skins of black grapes is quite different from that made by blending a little red into white wine. All European still pink wines are made directly from black grapes by removing the fermenting must from the skins after a short contact period, so that only small amounts of anthocyanins are extracted. The reason that pink champagne is made by blending red and white wines is that the main flavour comes from the champagne process rather than the base wine.

•   Short maceration

It is generally regarded that the best way of producing a pink wine is by the short maceration method, where the crushed mass of black grapes is macerated for such time as to produce the correct depth of colour and flavour, and the liquid is then drained off. The attributes of this method are that it is designed specifically for the production of pink wine as the prime product and not as an off-shoot from another process.

•   Saignée

This is another way of producing pink wine, named after the French term meaning 'bled'. Some of the fermenting must is bled off after a short period of maceration, which might amount to two or three days. This has the two-fold purpose of producing a pink wine from the portion that has been removed, leaving the remainder of the must in contact with the entire mass of skins and thus producing a more concentrated wine.

•   Vin d'une nuit

The French use this rather romantic term for a pale pink wine made by draining off the must after a single night on the skins.

•   Double pasta

This is a Spanish technique that has been used for decades and involves manipulating two vats of fermenting must. The must from the first vat is drained off after a few days and is finished as a *rosado,* with the skins being transferred to the second vat, which produces a thick, black red wine from its double dose of polyphenols. The main purpose of this operation was the production of the black wine, which could be sold in bulk for blending purposes. The *rosado* was originally merely a by-product, but nowadays is a very respectable dry wine and the equal of anything from Provence.

Chapter 9

WHITE WINE PRODUCTION

No longer drink only water, but use a little wine for your stomach's sake and your frequent infirmities.

1 Timothy 5.23

The production of white wines is relatively simple compared with the production of red wines because the must is separated from the skins before fermentation and thus there is no complication with floating solids. All the effort in making a good white wine goes into the preservation of the aromas by cool fermentation, and by the prevention of oxidation. In those instances where skin contact is practised (see p.104), the juice is usually separated from the skins before fermentation begins. Although the techniques are similar to those used for red wine production, the order of operations is different:

- de-stemming
- crushing
- sulphiting
- skin contact
- draining
- pressing
- clarification
- adjustments to acid and sugar
- fermentation

Some of these processes have certain conditions that are of particular importance in the production of white wines.

Cool fermentation

The ready availability of refrigeration equipment for cooling and temperature control has encouraged winemakers to ferment white musts at lower and lower temperatures, but unfortunately, excessively low temperatures do not produce attractive wine. The results of such techniques are wines that are clinically clean, with aromas reminiscent of pineapple and banana, yet lack true varietal character and taste very similar, whatever their provenance.

It has been realised that there is an optimum temperature, which depends upon the grape variety and the style of the wine being produced, but will probably lie between 15 and 20°C. The tendency now is to keep towards the higher end. At too low a temperature, too many of the volatile esters are retained, with the predominance of the estery pineapple aroma. As the temperature rises, so the true character of the grape is allowed to show itself. At high temperatures the volatile components are lost, and the wine becomes dull and lifeless.

Skin contact (macération pelliculaire)

Red wine, of course, gets its colour from the skins of black grapes, usually during the fermentation. For the sake of clarity of definition, this is usually referred to as maceration or skin fermentation. The expression skin contact, or *macération pelliculaire,* is reserved for the production of white wines and should not be confused with *macération carbonique*, which is used in the production of red wines (see p.97).

In the making of white wine it was (and still is in many instances) normal practice to separate the juice from the skins as quickly as possible, because it is recognised that the flavour could be spoiled by extraction of the harsh and bitter polyphenols in the skins. As the study of winemaking advanced, it was realised that many of the flavouring components of aromatic grapes are contained in the layer of cells under the skin, immediately adjacent to the pulp. After the grape has been picked, provided the skin is still in contact with the juice cells, these flavour compounds gradually leach into the juice. It is therefore necessary to allow a certain amount of time for this to happen, which is contrary to the accepted technique of rapid processing.

The usual method of skin contact, or *macération pelliculaire,* as this part of the process is known, is to crush the grapes carefully and then leave the crushed mass to stand for several hours with the skins in contact with the free run juice. Reducing the temperature minimises the risk of extracting unwanted flavours because these compounds are less soluble at lower temperatures. Keeping the mixture of skins and juice for as long as two days at zero degrees celsius (32°F) is sometimes done to bring out the maximum flavour, the low temperature preventing bitterness from developing.

This method is particularly useful for aromatic varieties such as sauvignon blanc, but it does not work for all grapes as there is a danger of extraction of polyphenols at the same time.

An alternative technique, requiring carefully gathered and undamaged bunches of grapes, involves simply leaving the grapes in a cool place overnight before crushing and pressing the next day. This somewhat subtle process results in a diffusion of the aroma compounds from the cells on the inside of the skins, while avoiding any contact with the exterior surface and minimising the release of polyphenols. It is essential that the grapes should be in as perfect a condition as possible to minimise damage due to oxidation.

Sometimes the berries are not crushed but whole bunches are loaded directly into the press. This method is imperative with high quality sparkling wine production, as used in Champagne. Whole-bunch pressing leads to a juice with fine, delicate flavours, low phenolics and low solids. Some hot-climate Chardonnays and Rieslings are made in this way.

Tank vs. barrel

Many modern white wines are fermented entirely in stainless steel tanks because this produces a clean, fresh wine as appreciated by the modern palate, and the tanks are easy to clean and sterilise. White wines can be finished in oak barrels if more complexity is wanted, but this is not the ultimate method. The greatest complexity is achieved by fermenting in barrel, as opposed to merely maturing in barrel. This is the result of multiple reactions between the polyphenols in the wine, in the wood of the barrel and in the yeasts themselves. (See ch.11 for greater detail.)

*Tumultuous fermentation
in a traditional
barrel fermented
white Burgundy*

Sur lie

This is a French term meaning "on the lees", the lees being the deposit in the bottom of a tank or barrel at the end of fermentation. It consists of yeast cells, both dead and viable, and particles of grape skin and cells. The purpose of this technique, which is used mostly for white wines, is to induce more flavour and greater complexity, and usually a slight 'toasty' quality. The wine usually associated with this technique is Muscadet, where the wine is left on the lees for several months. By law, it must not be racked and must not be bottled before the end of the March following the vintage.

Bâtonnage

An extension of lees contact, as above. When a wine is left in contact with a thick deposit of lees for several months, the dead yeast cells start to decompose under the action of their own enzymes. This creates what is known as a reductive condition, which means the opposite of oxidation. Under these conditions, some of the sulphur dioxide is reduced to hydrogen sulphide, a foul smelling compound (dirty drains, bad eggs). Stirring introduces oxygen to the lees, which prevents the reductive condition occurring, and thus prevents the creation of a foul smell.

Traditional 'batonnage' equipment   *Modern barrel rollers*

Stirring the lees also increases the contact of the wine with the dead yeast cells, thus enhancing the flavours produced by these cells.

Bâtonnage is normally carried out by inserting a stirring rod with a chain attached to the end through the bung-hole and stirring in a circular motion to agitate the deposit and mix it into the wine. A neat alternative, which is quicker, easier and more effective, is to stack the barrels on a set of rollers, so that the barrels can be agitated by rotating rather than stirring.

Prevention of oxidation

One of the modern principles of the making of white wine is the prevention of oxidation, for oxygen is the great destroyer of fruit. Red wines contain high levels of polyphenols which act as natural antioxidants so they are less susceptible to oxidation, while white wines do not have this protection. The old style white wines from Spain, France, Italy and eastern Europe were dreadfully dull, brownish in colour, with a nose of wet cardboard and a palate tasting of anything but fruit, all rounded off with a whiff of sulphur dioxide. This was mostly the result of poor oxygen control.

One of the revolutions in modern winemaking has been the production of pale coloured, delicately fruity white wines, refreshing and eminently drinkable. Much of this style is the result of the elimination of oxygen at every stage in the winemaking process:

- Pressing the grapes in a tank press, pre-flushed with nitrogen

- Moving the juice through pipework and into vats where all the air has been removed by flushing with carbon dioxide or nitrogen

- Checking all joints for integrity of seals, especially pump seals

- Never keeping wine in a part-filled tank unless blanketed with nitrogen

- The correct use of antioxidants, especially sulphur dioxide and ascorbic acid

- Attention paid at all times to keeping dissolved oxygen at low levels

Sweet wines

The sweet wines of the world are made by various methods according to local traditions - and cost. They can be classified as follows:

- The addition of concentrated grape must to a fully fermented dry wine (cheap carafe wine)

- The addition of preserved grape juice to a fully fermented dry wine (German QbA wines)

- Using grapes that have become over-ripe and naturally concentrated by the sun (the so-called *cuvee* wines and some German QmP wines)

- Using grapes that have been picked when ripe and dried by keeping the bunches in a dry atmosphere (Amarone and Vin Santo)

- Using grapes that have been affected by *Botrytis cinerea* (Tokaji Aszu, Sauternes and German *Trockenbeerenauslesen* wines)

- Using grapes that have been frozen (German *Eiswein* and Canadian Icewine).

Grapes drying for Vin Santo at Villa di Vetrice in Pontassieve

• Carafe wines

The simplest and cheapest way is to add some grape sugars to dry wine that has finished its fermentation and has been clarified and stabilised.

It should be noted that the sweetening has to be done using grape must in some form; it is not permitted to use sucrose (except in the case of Champagne). This is the method used in the production of low cost carafe wines, such as would be called a "Medium Dry White Wine" or a "Sweet White Wine" as served in many pubs. Actually, some low cost red wines contain a low level of added sugar to produce a style that is preferred by the everyday palate.

• *German wines*

Dry wines have been made in the southern parts of the German wine region for centuries. These have sufficient body to make well-balanced wines, suitable for drinking with food. In the more northerly sectors such as Rheingau, Rheinhessen and Mosel, the wines are of a lighter weight and have traditionally been made in a sweeter style by the addition of *Süssreserve*, which is unfermented grape juice that has been preserved by various methods such as microfiltration, cold storage or the addition of sulphur dioxide.

This addition of sweetening is allowed only for the production of *Qualitätswein bestimmte Anbaugebiete (QbA)* which is the classification used for the lower end of the quality scale. It is not allowed for the higher quality *Qualitätswein mit Prädikat (QmP)*, where all of the residual sugar is natural to the original grapes. These

Riesling vines in the Piesporter Goldtröpfchen vineyard

wines can be classified into different categories according to the sugar levels in the grape juice from which they are made. The following figures are for Riesling wines produced in the Rheingau:

Kabinett	73°Oe
Spätlese	85°Oe
Auslese	95°Oe
Beerenauslese, Eiswein	125°Oe
Trockenbeerenauslese	150°Oe

(See p.231 for explanation of °Oe)

With the modern fashion for dry wines, many producers have been experimenting with fermenting some of these wines to dryness, but the result is not always successful - and it makes it very confusing to discover that a wine labelled *Auslese* is dry! The fact that the German wine producers use several of the various sweet wine methods has resulted in their incredible range of tastes, which are some of the finest sweet wines in the world, with wonderful delicate and complex aromas and fruits balanced by a fine acidity.

(Note that *süssreserve* is not used for the process of enrichment before fermentation to increase the alcoholic content of the finished wine.)

•   *Sauternes*

Sauternes is the epitome of wine produced from grapes that have been attacked by *Botrytis cinerea* or *pourriture noble*, otherwise known by the less attractive name of Noble Rot. The development of this condition depends entirely on the right weather conditions. It is necessary to have misty, moist mornings that encourage the growth of the fungus, followed by sunny dry afternoons to halt its growth.

It follows, therefore, that the production of the finest wines only occurs in those years when the conditions for the development of the fungus is right. In a good year, when the berries have shrivelled, each one has to be picked by hand, leaving the healthy berries to succumb to the noble rot in due time. This means that the pickers have to pass through the vineyard several times in order to pick the berries that are in the correct condition. It is not unusual in a vineyard such as Chateau d'Yquem for the harvesters to go through the vines eight or nine times, hence the high cost of wines such as these!

• *Tokaji Aszú*

The *aszú* wines of Tokaj are made by a unique method that is not used in other wine region, so it is worth looking at them in detail.

As with botrytised wines produced in any region, the weather plays a very important part in their quality, depending on damp, misty mornings and dry, sunny afternoons. The shrivelled berries are picked individually by hand, a very tedious and exacting task, usually done by the deft fingers of women. At this point the process is totally different.

First quality aszu berries selected by Royal Tokaji for the production of Tokaji Aszu

The botrytised berries are collected in bins at the winery and are stored after being dusted with potassium metabisulphite as a preservative. The healthy berries are then harvested and used for making a normal dry white wine.

The next step is to give the *aszú* berries a gentle crushing and add them in measured quantities to the dry base wine. After macerating with careful mixing for a few hours, the wine is drained off the skins and is transferred to barrels in the deep, cool cellars where it stays for a minimum of two years, but often much longer.

During this period the wine gradually matures, sometimes with a very slight fermentation, producing an incredibly complex, rich and sweet wine, with a characteristically clean and sharp finish. This acidity which balances the lush sweetness is due to the particular grape varieties used, especially the Furmint. The other varieties, the Hárslevelu and the Sárgamuskatály add to the complexity.

The deep, cold cellars of Royal Tokaji

The classification of Tokaji wines is not easy to understand, partly because of the unique character of the Magyar language. It is actually quite simple and is linked to the sugar content of the finished wine, unlike German wines, where the categories are determined by the sweetness of the unfermented must. The name originates from the word for the hod in which the grapes used to be collected: *puttony* or *puttonyos* (plural).

3 puttonyos = 60 - 90 g sugar per litre
4 puttonyos = 90 - 120 g sugar per litre
5 puttonyos = 120 - 150 g sugar per litre
6 puttonyos = 150 - 180 g sugar per litre

The juice that drips from the *aszú* berries during storage is the famed *Eszencia* and is more akin to fruit juice than wine, as it contains anything up to 600 g/l of sugar and only 3% alcohol.

Chapter 10
SPARKLING & FORTIFIED PROCESSES

Burgundy for kings, champagne for duchesses, claret for gentlemen.

Anon French Proverb

Sparkling wines

Sparkling wines can be produced by various methods ranging from the most expensive, where the bubbles are produced by a second fermentation in the bottle, to the cheapest, where the gas is injected into the wine in the tank by an industrial method. The quality is reflected by and large (but not necessarily) in the price, because the best method involves a tremendous amount of manipulation.

•   *Traditional method*

This is the method that used to be known as the *méthode champenoise* because all champagne has to be made by this method. Such is the determination of the *Champenois* to protect their product that the rest of the world has to describe this method as the *méthode traditionelle* or the traditional method. This relies on the creation of carbon dioxide inside the bottle produced during a second alcoholic fermentation, and is the method most highly regarded for the production of a quality sparkling wine.

The first step is to make a dry base wine exactly as would be done for a still light wine, except that sulphur dioxide is not added at the end of

*Bottles lying sur lattes
with traditional agraffes
at Alfred Gratien
(see over)*

the fermentation. When the time comes for bottling, it is blended with *liqueur de tirage*, wine that has been sweetened with cane or beet sugar (sucrose). A selected yeast is added and the bottle is closed either with a crown cap or, more traditionally, with a cork held in place by a metal clip known as an *agraffe*. The bottles are then stored in the cold cellars hewn into the chalk of the Champagne region, lying on their sides with the stacks of bottles stabilised by lathes of wood, hence the expression *sur lattes*.

Deposit on the side of the bottle

The wine re-ferments during this period in the cold cellars, producing carbon dioxide which cannot escape. Unfortunately, it also creates a deposit of yeast that would render the wine cloudy on opening if not removed. Conversely, it is this deposit that gives these wines their character during the long period of maturation on the lees in the bottle, when the yeast cells die and gradually decompose (autolysis). Much of the complexity of champagne is developed during this period, but the quality of the original wine, the *vin clair*, does play a considerable role because the poorer wines would not survive this long period in bottle.

When the time comes for the wine to be sold, the deposit in the bottles has to be removed, otherwise the wine would be cloudy when opened. This is done traditionally by the manual process of *remuage*, where the bottles are placed in specially designed racks known as *pupitres*, and a *remueur* goes round every day giving each

Bottles in traditional pupitres

bottle a sharp twist and a move towards the vertical position. This process gradually moves the deposit into the neck of the bottle.

But this is a tedious and expensive process, so it is now being carried out more and more (even in Champagne) by the automatic riddling machines known as *gyropallets* in France, *gyrasols* in Spain and *VLMs (Very Large Machines)* in the USA. These machines make a sharp quarter turn every thirty minutes which has the same effect as the old *remuage,* but the result is achieved more quickly and efficiently.

At the end of this process the deposit will be sitting either on the cork, if the original system is being used, or in a small plastic collection cup known as a *bidule* if a crown cap has been used. The bottles are now standing vertically upside-down, or *sur points,* and are transported in this position to the bottling line.

The generally accepted procedure nowadays is to freeze the neck of the bottle to trap the deposit in a plug of ice. When the cap or cork is removed, the ice shoots out of the bottle, taking the deposit with it. It is worth noting that the bottle in which this wine is sold is the actual bottle that has held the wine during its long period of second fermentation.

The bottles are then topped up with wine and a measured quantity of *liqueur d'expedition*, a solution of sucrose in wine that gives the finished champagne the correct balance. (The popular Brut style contains up to 15 g/l.) All that is then required is a new cork, held in place

Crown cap and ice plug

with a wire cage known as a *muselet,* a capsule and a label, and the wine is ready for sale.

Champagne is the obvious example of wines made by the *méthode traditionelle,* but all English sparkling wine is made by this method, as is Cava in Spain. There are many wine producing countries around the world making good quality sparkling wines by this method although, as stated above, the ultimate quality depends mainly on two parameters: the quality of the original base wine and the length of maturation on the lees before disgorging, which is a minimum of fifteen months for non-vintage champagne, and three years for a vintage.

•   *Transfer method*

This is another approach to the problem of ridding the sparkling wine of its yeast sediment. It is undeniable that the best flavours in sparkling wine arise from prolonged contact with the yeast deposit, especially when in close contact, as occurs in individual bottles. The result is never as good when the second fermentation takes place in a bulk tank (see below).

In the transfer method, the second fermentation takes place in the bottle, as for the traditional method. The difference lies in the removal of the deposit. In this case, the bottles are emptied under carbon dioxide pressure into a pressurised tank where the wine is cooled to -5°C. After the addition of the *dosage* (the sweetening wine), the wine is filtered and bottled.

The result is a wine that has the flavour characteristics of bottle fermentation, but with a slight loss of quality due to the handling it receives during the transfer process.

•   *Tank method (Cuve Close or Charmat)*

In this method, the second fermentation takes place in a pressure tank rather than in individual bottles. The greatest shortcoming of this process is the lack of contact between the yeast deposit and the wine itself. Some producers try to compensate for this by stirring the lees periodically, which does introduce a degree of complexity.

However, the basic problem is one of time. It would not be practicable to hold tanks of wine for the number of years it takes to produce the same effect as bottle fermentation. Nevertheless, some excellent sparkling wines are produced by the *cuve close* method, and at very good prices.

• Carbonation ("Pompe bicyclette")

In this method, the carbon dioxide gas is not produced by any form of second fermentation but is introduced from a commercial source of gas supplied in compressed form. The wine is first chilled, and carbon dioxide is bubbled into it, when it will readily dissolve. There are none of the extra nuances of flavour produced by fermentation and the gas gushes out again as soon as the pressure is released – a very inferior method!

• The Asti method

This was developed for the production of what used to be called Asti Spumante, now known simply as Asti. Its purpose was to preserve the flowery characteristics of the Moscato variety which disappear if the wine is fermented out to dryness.

The Moscato must is pumped into a pressure vessel and a yeast culture added. Fermentation occurs and the carbon dioxide generated is allowed to escape to atmosphere. When the alcohol level reaches around 5% vol, the valves are closed and the subsequently produced carbon dioxide becomes trapped in the wine, rendering it sparkling. At a pre-determined level of alcohol and sugar (usually 6-9% alcohol and 60-100 g/l of sugar), the contents of the tank are cooled to 0°C which stops the fermentation. The wine is then clarified, filtered and bottled.

Bottling poses a major problem with Asti. It is very susceptible to re-fermentation in the bottle by virtue of the fact that the alcoholic strength is around 9% and the residual sugars are about 35g/l, which presents the yeasts with a very nutritious medium in which to flourish. One producer who was asked why he uses both membrane filtration and tunnel pasteurisation replied, "So I can sleep at night!"

Fortified wines (liqueur wines)

According to EU legislation, what used to be known as fortified wines are now liqueur wines. The practice common to them all is that they have alcohol added beyond that produced by the fermentation of the original grape must.

•   *Vins doux naturels (VDN)*

The title of this category of wine is misleading as they are not naturally sweet, but are only sweet because of the addition of alcohol that prevents further fermentation. It could be said that they are naturally sweet in that all of the sugars in the finished wine were present in the grapes from which the wine was made. Nothing has been added apart from the alcohol to stop the fermentation.

The *vins doux naturels* are simply musts that are partially fermented, with the fermentation stopped by the addition of alcohol to bring the level up to between 15 and 18% vol. Because of the partial fermentation, the wines contain residual sugar and are sweet.

Subsequent treatment varies considerably according to the style of the ultimate product. Muscat wines are stored anaerobically in tanks; others are matured in oak barrels. Rancio wines are deliberately oxidised by being stored in large wooden vats over a period of years, with regular refreshing by removing part of the contents and replacing with young wine.

Wines of this style are exemplified by the delicious, unctuous Muscats from the south of France: *Muscat de Lunel, Mireval, Frontignan, Saint Jean de Minervois* and *Rivesaltes.* These are often drunk as an aperitif by the French, who prefer sweet wines at this stage in a meal.

•   *Port*

The basic principle behind the production of port is similar to that of *vins doux naturelles* in that fermentation is allowed to commence, but is then stopped, or muted, by the addition of alcohol. The differences are, first, that the grapes must be grown within the Upper Douro delimited region, and second, the grapes are treated differently during the maceration process.

The particularly interesting aspect of port production is the use of the *lagar*, or stone trough in which the grapes are trodden. The dimensions are approximately four metres square by less than one metre in depth, which are most unusual proportions for a fermentation vessel.

Upon arrival at the winery, the grapes may or may not be de-stemmed and are tipped into the *lagares* where they are crushed by the bare feet of a band of, usually, men. This is an arduous process and takes several hours to complete. The effect is to split the grapes open, release the juice and give a certain amount of abrasion to the skins, which releases the polyphenols. The shallow format of the *lagar* gives good contact between skins and juice, and also allows oxygen to dissolve readily, an important factor in the stabilisation of the anthocyanins.

The tedium of this process was greatly relieved by the introduction of the autovinifier, or Ducellier vat (see p.95), but the result was not as good. The action of the human foot obviously has a particular effect, which stirred the creativity of the port winemakers, who reverted to early practice and designed the 'mechanical lagar', a machine which closely mimics the action of treading. This has now become the preferred method of vinification for high quality port.

Fully automated robotic lagar for the production of Graham's Port

The mechanical, or robotic, lagar consists of a stainless steel trough of similar dimensions to the traditional lagar. Above this trough is the machine that houses the automatic 'feet' composed of rectangular pistons fitted with silicone rubber pads of a similar texture to the human foot. This machine runs across the lagar on rails, the pistons reciprocating in a vertical dimension, just like a row of human feet. This action squeezes the grapes against the bottom of the lagar, releasing the juice and squashing the skins. At a later stage in the process, the feet can be adjusted so that they move only halfway down the lagar, submerging the cap as in a *pigeage* operation and at the same time picking up oxygen from the atmosphere, giving a micro-oxygenation to the fermenting must.

After a short fermentation in the lagar the must is transferred to the traditional pipes, which are wooden casks of 550 litres capacity, where it is fortified by adding brandy to bring the alcohol to around 20% vol. This arrests the fermentation and thus retains the sugars to produce a sweet wine.

•   *Port styles*

After several months the wine is classified and the best is reserved for the production of **Vintage Port**. This wine is kept in full casks and must be bottled within two years of the vintage. It usually needs another 10 to 15 years in bottle to reach full maturity, after which it needs decanting from the thick sediment before drinking.

Port vintages are 'declared' only in the best years. In those which are not declared as vintage years, wine from a **Single Quinta** are often bottled as vintage port under the name of the vineyard or 'quinta'. These receive the same treatment as vintage ports and are often of good value, as they cannot attain the same price level.

Coming down another level of quality is the **Late Bottled Vintage**, which is bottled after five or six years in wood. This wine will have deposited all of its excess polyphenols in the barrel and will not need decanting before drinking.

Tawny Port has remained in wood for a much longer period and is classified according to the length of time in the wood. The standard periods are 10, 20, 30 and 40 years. The longer the period, the more complex and lighter they become. Cheap Tawny is a blend of Ruby and White Port and is of a much inferior quality.

Colheita Port is a tawny port of a single vintage and has a character similar to an aged tawny.

Ruby and **Vintage Character** are the lower end of the quality spectrum, and are aged in wood for a short time before being bottled and put on the market for immediate drinking. They do not require decanting.

White Port is made only from the white varieties and is much admired in Portugal as an aperitif because it is vinified with a lower sugar content than the red styles. However, it oxidises readily and should be drunk young.

An interesting chemical fact is that the natural iron component in ruby and tawny ports is in the ferric state, which indicates aerobic maturation. These wines do not improve in bottle. The iron in vintage port is in the ferrous state, indicative of anaerobic storage and also of the potential for development in bottle.

• *Sherry*

The particular characteristic of sherry production is the solera system and the phenomenon of *flor* which is a skin-like yeast growth that forms on the surface of some of the wines that are maturing in the butts (barrels).

The *solera* system consists of a series of barrels arranged in rows, one on top of the other, usually four high. The principle of the system is that the new wine goes into a row of empty barrels, known as the *añada*, meaning the wine of the year. At the other end of the *solera* the last row of barrels contains the oldest wine ready for bottling. This row is actually called the *solera* row, although the whole system is also known as the *solera* (very confusing!).

The idea is that when wine is required for bottling, it is drawn from the *solera* row, but never more than one third of the contents at any one time. These barrels are then topped up from the next oldest row, and they are topped up from the previous row and so on, until the youngest barrels are reached. These intermediate rows, or scales, are known as the *criadera*, meaning nursery, or where the wine is nurtured. The finest finos, such as Tio Pepe from Gonzalez Byass or Domecq's La Ina, can have as many as eight scales.

The purpose of this complicated arrangement is multiple blending, but with younger wines always added in smaller quantities to older wine, so that the older wine has the greater influence. By this system it becomes obvious that there is normally no such thing as vintage sherry. If a date is seen on the label of a bottle of sherry it will be the year of establishment of the solera and this will usually be many years in the past. In theory, there will still be a few molecules of the original wine remaining in the blend.

Sherry solera
at
Sanchez Romate

The grapes, mostly Palomino Fino, are grown, harvested and fermented in stainless steel tanks like any other white wine, but at a higher temperature because the aim is not to produce a fruit driven wine, but something rather bland. At the end of fermentation the wine is stored in as the *añada*, where it stays for a few months. The barrels are then tasted individually in order to decide whether they are heading towards a fino or an oloroso style. This having been done, they are transferred to the first scale of either a fino or an oloroso *criadera* as appropriate.

The **fino** wines are fortified to between 15 and 15.5%, but no higher, so that the mysterious flor yeast can develop on the surface of the wine. This film consists of yeasts of various strains of *Saccharomyces* that form a continuous layer over the surface, protecting it from oxidation and at the same time forming acetaldehyde and other substances that produce the typical fino aroma and flavour. Regular extraction of wine from the *solera* and the topping up with young wine are important factors in the maintenance of the fresh character of fino.

Those butts where the flor does not grow adequately are fortified more strongly to around 17.5% to destroy any remnants of the flor, and are moved into an **oloroso** *criadera*. In these butts, oxidative conditions prevail and the wine becomes darker in colour and more fragrant, hence the name *oloroso*, meaning fragrant.

This explains two of the main styles of sherry, leaving **amontillado**. This is an aged fino, where the flor has gradually died out, leaving the fino to oxidise and become darker in colour and more and more intense with age. However, the constant rejuvenation with younger wine is still important to prevent total oxidation.

Some of the romance of winemaking has been lost in our modern scientific age, and this is particularly apt in the production of sherry. No longer do we have to wait expectantly for butts of sherry to turn mysteriously one way or the other, towards fino or oloroso. Knowledge of microbiology has enabled us to force the development whichever way is commercially necessary. In many *bodegas*, fino is produced in a bulk tank by inoculating with a flor yeast and bubbling air through it to supply the yeast with the oxygen that is necessary for the formation of the aroma compounds. When the flor has developed sufficiently, the wine is put into a fino *criadera*.

It should be noted that all sherry is dry when it is first drawn from the solera, because all the sugar will have been converted to alcohol. The very best old amontillados and olorosos are bottled without further blending and are labelled as *viejo*, meaning old. These are wonderful, intense, complex wines. Although expensive, they are drunk only in small portions, so are are good value for money and well worth searching for.

Equally delicious are the true sweet wines made from the Pedro Ximenez grape which is high in sugar, picked at full ripeness and allowed to dry in the hot sun of Jerez. These wines are black in colour, rich and ripe and totally unctuous - fabulous!

The standard range of medium-sweet wines are blends of lesser amontillados and olorosos, sweetened and coloured with wines specially prepared for the purpose.

• *Madeira*

The different styles of Madeira are produced by following either the port or sherry systems: sweet styles by the port method of adding brandy to halt the fermentation, the drier ones by allowing the yeasts to complete the fermentation, as with sherry.

The particularity of Madeira is the use of the *estufagem* process, by which the wine is heated very slowly over several weeks to between 40 and 50°C. After this treatment it goes into a solera system as with sherry.

The various styles were originally named after the grape varieties from which they were made, being, from the driest to the sweetest: **Sercial**, **Bual**, **Verdelho** and **Malmsey** (or Malvasia). After the ravages of Phylloxera, a somewhat relaxed attitude prevailed and the wines were produced from Tinta Negra Mole and Complexa. Since joining the European Union, the wines have to be made from a minimum 85% of the named variety if they are to be so labelled.

• *Marsala*

Produced around the town of Marsala on the western tip of Sicily. It is made by the port method, with the addition of brandy according to the desired sweetness of the resulting wine: early in the fermentation for sweet wines, and progressively later for the drier styles. It is finished in a solera system, as with sherry.

It is classified according to the minimum length of time in the solera:

Marsala Fine - 1 year
Marsala Superiore - 2 years
Marsala Superiore Riserva - 4 years
Marsala Vergine - 5 years

Marsala has acquired the reputation for being a wine merely for cooking, which is very unfair because the best aged Marsalas are excellent fortified wines.

It is worth noting that very good light wines are now being made on Sicily, using modern techniques of temperature and oxygen control.

Chapter 11

WOOD & MATURATION

Many strokes, though with a little axe,
Hews down and fells the hardest-timbered oak.

Shakespeare, Henry VI Part Three

The influence of wood on wine is a complex and controversial subject. It is undoubtedly ancient in origin, as wooden vessels were in use long before concrete, fibreglass or stainless steel. Conversely, one must not forget that other materials were in use long before the technology of wood-working had been learned: animal skins, earthenware vessels, stone jars and troughs.

Oxygen plays an important role in the maturation of red wines in wood. Much of the character of fine red Bordeaux wines, for example, depends on the period of time they spend in oak barrels (Fr. *barriques).* During their sojourn in the barrels, the oxygen that gets in through the pores of the wood, around the bung and during the process of racking, attacks the tough tannins in the wine, causing them to precipitate and fall to the bottom of the barrels, leaving the wine softer and more supple. (See also p.133 Micro-oxygenation)

Oak trees in a French forest

Oxygen also plays an important role in the stabilisation of the colours in red wine. At the end of the primary fermentation, oxygen is needed in the creation of stable colouring matter to enable the tannins and the anthocyanins to interact by a process known as 'oxidative coupling'. This ultimately produces more stable colouring matter.

A secondary, and some would say the prime, purpose of using wood is the addition of extra flavours and therefore the increase in complexity of the wine. The degree to which these flavours are imparted to the wine depends on the number of times the barrel has been used, on the length of time the wine spends in the barrel and the size of the barrel.

New barrels have to be used if maximum influence is required, and it is usual to find that the wine is in the barrel for only some eight months. New barrels are expensive, in the region of €500 each, so it is not surprising that wine that has been through a proper regime of fermenting and ageing in wood is also expensive.

Type of wood

There is a tendency to think of oak as the only wood used for making wine containers, but this is a somewhat narrow view as other woods such as chestnut, beech, cherry, acacia, walnut and mahogany have been used. Cherry is used in the Veneto for the ageing of Valpolicella Ripasso because it softens the polyphenols without adding any vanilla notes.

Acacia finds its followers in Veneto and in Austria where it is appreciated because it has soft tannins and a sweet neutral taste for the maturation of Sauvignon Blanc, Grüner Veltliner and Riesling. The wood used is from *Robinia pseudoacacia* and seems to be finding its place for the maturation of aromatic white varieties.

However, there is nothing to equal oak for its combined properties of storage and flavour. In fact, one might almost put flavour before storage, such is the demand for the characteristic nuances of vanilla and spice that good oak can bestow.

The correct use of oak is complex and involves many decisions, so it is worth looking into this topic more deeply.

Oak

Oaks are members of a large family, the *Fagaceae*, which is actually the beech family. Within this family, the oaks belong to the genus *Quercus*, of which there are over 250 species, but only three of these are of any interest for the construction of barrels.

The most basic subdivision is between European oak and American oak. European oak is found in two distinct species, *Quercus robur,* the English oak or pedunculate oak, and *Quercus petraea*, sessile oak, which used to be known as *Quercus sessilis*, the oak found in most of the French forests. American oak is *Quercus alba*, the best of which is found in Pennsylvania, Minnesota and Wisconsin.

Good European oak suitable for barrel production is produced in several countries, including Hungary, Romania, Russia and Poland, but France is probably regarded as the best source, if only because it has the largest reserves in Europe and the most varied. *Quercus petraea* is generally regarded as the finest oak because it has a tight grain and is rich in aromatic compounds. This is the species found in the Tronçais, Nevers, Allier and Vosges forests. Limousin is the only forest composed of *Quercus robur*, which has coarser grain and a lower aromatic content and is particularly useful for the ageing of cognac.

Coopers, however, are not particularly concerned about the species, but regard the geographical origin as more important. Hungarian oak is said to add a sensation of weight and texture, whereas Caucasian oak gives a moderate tannin and aromatic notes. One of the most favoured regions is Slavonia (not to be confused with Slovenia) which is a part of Croatia, where the oaks are *Quercus robur*. The coarser grain structure gives the wine an increased micro-oxygenation and helps to integrate the various elements of the wine as well as giving it a longer life in the bottle. It is particularly used by growers in the Veneto, Toscana and Piemonte regions of Italy.

It has been reported that a Spanish cooper is using Chinese oak, *Quercus mongolicus*.

Oak maturation vats at Ogier in Châteauneuf-du-Pâpe

Size of vessel

It must not be forgotten that oak is used not only for barrels, but also for large vats, for both storage and for fermentation. These upright vats are normally made in sizes between 10 and 200 hl, but can sometimes be found at 500 hl or even larger. The so-called "largest oak vat in the world" can be found in the town of Thuir in southern France, holding one million litres (10,000 hl).

Large oak vats are used not for the benefit of oak flavour but for other reasons:

- good thermal insulation;
- oak tannins (ellagitannins) at a low level assist in the structuring of the polyphenols;
- low level micro-oxygenation assists in stabilising the colour of red wines;
- they look good!

Considering barrel sizes, the Bordeaux *barrique* of 225 litres seems to be regarded as the optimum capacity to give a sensible oak influence with ease of operation. The rate of change in a liquid always follows the same rule, that it is inversely proportional to the size of the container. This is because the changes occur at the interface between the liquid and the walls of the container, and the smaller the container, the greater the ratio of surface area to volume. Even in Hungary, where the Tokaji Aszu producers have traditionally used the 130 litre *gönci*

Bordeaux wine maturing in barriques at Château de la Grave, Côtes de Bourg

barrel, the *barrique* seems to be gaining favour.

The choice of barrel size is important in relation to fermentation temperature. Smaller barrels, which have a greater surface to volume ratio than larger barrels, are more efficient at dissipating the heat of fermentation. This is the reason why the traditional size of 225 litres was adopted in Bordeaux, being appropriate to fermentation in a cool cellar. Refrigeration is rarely needed, warming being more usual to prevent a fermentation stopping prematurely, where wines are still fermenting in the late autumn in a particularly cold season.

Seasoning and toasting

All oak has to be seasoned before use. This occurs after splitting into

stave-sized pieces, when the moisture content is around 55%. After three years this will have fallen to about 15%. In common with all other forms of joinery, kiln drying is sometimes used but this does not give a good result, giving astringency and green characters to the wine from the rough tannins that are cooked into the wood.

Staves weathering at Château Bel Air in Saint-Emilion

It is being realised more and more that the natural seasoning process is not only critical but complex. Recent research has shown that microorganisms play a large part in preparing the wood to receive wine. These organisms grow deeply into the pores of the wood causing chemical changes in the polyphenols that result in adding elegance, richness and complexity to the wine, with a smoother finish. Artificial kilning destroys these microoganisms and the benefits they bring are lost.

All barrels have to be heated in order to make the staves more flexible, so that they can be bent into shape. This takes about 20 minutes, and the temperature of the

Firing a barrel at Taransaud

wood reaches between 120 and 180°C. This results in a light toasting level. After another 10 minutes or so, the temperature will have risen to about 200°C and gives a medium toast. A further 5 minutes produces a temperature of 225°C and a heavy toast. At this stage the interior of the barrel is deep black.

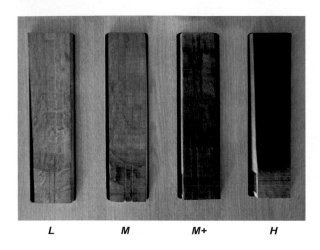

The results of toasting ranging from light through medium and medium plus to heavy

L   M   M+   H

Toasting is a huge topic, with much research still in progress. To put it in basic terms, the greater the degree of toasting, the greater the influence on the flavour of the wine.

The resultant aromatic substances are many and complex. From the lignin in the wood comes the phenolic aldehyde vanillin, probably the best known and best loved added flavour. Other compounds produce aromas described as "smoky, roasted, spice, clove, carnation etc."

Compounds derived from the hemi-cellulose in the wood produce yet more aromas: "toasted bread, caramel, cocoa, coffee, sweet pepper, roasted almond etc."

Fermentation in barrel

There has been an upsurge in barrel fermentation in recent years because it has been realised that better integration of oak flavours can be achieved by fermenting in the barrel, as opposed to merely ageing in the wood. This technique can be applied to red wines as well as white, although in the case of red the initial fermentation during the maceration has to be conducted in a vat in order to manage the flotation of the skins. Only after the must has been separated from the skins can the liquid be transferred to barrels for the completion of the fermentation.

An advantage of barrels over tanks is that an ample supply of oxygen is available for the first stage of fermentation, enabling the yeast to reproduce more rapidly because it is in its aerobic phase. This gives a quick start to the fermentation and usually results in a higher alcohol in the finished wine.

The reactions going on in the barrel during fermentation are complex and have a major effect on the structure of the finished wine. The presence of yeast inside the barrel changes the interaction between wine and wood and plays an active role in the balance between fruit flavours and wood flavours. In the first instance, the yeast cells coat the inside of the barrel which reduces contact between wine and wood, and secondly, the yeast can absorb some of the wood extracts, changing them by a biochemical transformation into less aromatic compounds. The overall result is a more subtle and harmonious blend. Furthermore, the mass of yeast cells acts in a similar way to the polyphenols in red wine, protecting the wine from oxidation.

At the end of the fermentation there is the choice between racking the wine off the sediment or leaving the wine on the lees. In the latter case it is important that the lees are stirred periodically (Fr. *bâtonnage*), to prevent the risk of reductive flavours developing within the sediment. The mass of yeast cells has several further useful properties. As the yeasts die they release valuable substances known as polysaccharides, which act as natural fining agents, reducing bitterness from the wood tannins and thereby softening the wood flavours. They also increase complexity, improve the clarity of the wine, and prevent oxidation and a darkening of colour.

The disadvantages of barrel fermentation relate to the cost of barrels, the difficulty of hygiene and the large amount of manpower involved in the handling of these comparatively small units. There is also a loss of freshness and of fruit, and an acceleration of maturity. So this mode of fermentation is only used for those white wines which have sufficient body and structure and are valued for their richness and complexity, and is not suitable for young fruity wines for early drinking.

Maturation in wood

Fermentation in a tank, with transfer to barrel after the completion of fermentation, does not produce quite the same effect. The various inter-reactions are less complex and the result is less subtle. The extraction of the many flavour-producing compounds from the toasted wood is more pronounced in the absence of fermenting yeasts. The oxidation of the polyphenols is also more rapid, since the protective effect of the fermentation process is absent.

Putting the wood in the wine

If all that is wanted is the extra flavours from the wood and not the oxidative maturation imparted by an expensive barrel, then rather than putting the wine into wood, the wood can be put into the wine in the form of chips *(Quercus fragmentus?)*.

"Pieces of oak wood" according to EU nomenclature otherwise known as OAK CHIPS

It is possible to buy chips of all the standard oaks, in all degrees of toasting, so the winemaker can reproduce the flavour profile of a barrel without the expense.

But – the result is not the same. Obviously, the softening of the harsh tannins does not occur because there is no contact with oxygen. The oak flavours will be extracted, but they do not integrate in the same way as from a barrel.

Micro-oxygenation

There is a relatively new technique for achieving balance in red wine without the expense of oak barrels and it is gaining ground rapidly. The principle behind it is simple: if maturation in wood is the result of a very slow ingress of oxygen, then why not pass a very slow stream of oxygen into the wine in a tank? This would render expensive barrels obsolete. And it works!

The dose of oxygen required is extremely small, in the region of 1ml of oxygen per litre of wine per month. So, what is needed is a means of breaking the flow of oxygen into minute bubbles, bubbles which are so small that you actually cannot see them. This is easily done by forcing the gas through a cylinder of porous pot. The small pores of the unglazed pot cause the oxygen to diffuse into the wine rather than producing bubbles. It is most important that the rate of introduction of oxygen is almost at a molecular level rather than in the form of gaseous quantities.

A micro-oxygenator as used at Cave de Tain in l'ermitage

The action of oxygen on the wine is complex and depends on various factors such as the phenolic structure of the wine, the level of sulphur dioxide, the temperature and the timing. Although there is still some discussion regarding the way in which it works, there is no doubt that the main effect is to change the polyphenolic structure of the wine. The result is a wine that is smoother and softer, with better integration of the tannins, and yet will still develop in bottle. As might be expected, the best wines for this treatment are those which are high in both tannins and anthocyanins.

The micro-oxygenation (microx) treatment falls into two main phases:

1. The first dose of oxygen takes place at the end of the primary fermentation, probably before the onset of the malo-lactic fermentation. Quite high levels of oxygen are added: 10 to 60mg/litre/month, for 1 - 3 months. During this period, known as the **structuring phase**, changes occur in the structure of anthocyanins and tannins, as oxidation occurs, stabilising the colours of red wine. This reaction with oxygen is important in the production of all red wine (see p.25). These changes in the polyphenol structure cause the taste of the wine to become more harsh.

2. After the completion of the malo-lactic fermentation the second phase of microx can be applied. It is only during this **harmonisation phase** that the wine takes on the softness that is associated with microx. Throughout this period, generally lasting several months, the oxygen is added at a much lower rate, between 0.1 and 10mg/litre/month.

Microx is undoubtedly one of the most useful of modern developments and is being widely used. Its proponents claim that the benefits of microx are primarily an increase in the quality of the wine rather than an economic factor. It has long been recognised that oxygen plays a large part in the maturation of red wine; microx enables it to be done in a controlled fashion.

(N.B. Do not confuse micro-oxygenation with hyperoxidation, q.v. page 62.)

Chapter 12

PRINCIPAL COMPONENTS OF WINE

One barrel of wine can work more miracles than a church full of saints.

Italian proverb

The style and balance of each individual wine is dependent on the relative abundance of constituents originating from two different sources. The first group can be found in the original grape juice and the others arise as a result of the fermentation process. The grape juice supplies the sugars, the acids, the minerals and nutrients, the polyphenols which embrace the tannins for 'grip' and anthocyanins for colour, and the crucial flavouring substances (see ch. 3).

The fermentation creates other components that differentiate wine from mere grape juice. These include the alcohols, an expanded range of acids, glycerol and all the products of inter-reaction between these constituents.

Alcohols

During the process of fermentation yeasts produce alcohols, of which there are many different types. The predominant alcohol, by a considerable margin, is ethyl alcohol, or ethanol. This is the substance that has become abbreviated to 'alcohol' in all references to alcoholic beverages, conferring upon them their well-known characteristics: a warmth in the throat, a feeling of elation, then depression and ultimately drunkenness.

The ethanol content of wine plays an important role in the taste (as opposed to the flavour) of wine on the palate. Attempts to produce a non-alcoholic wine have not been exactly successful, despite utilising the best and most modern technique for removing alcohol, simply because the presence of alcohol has a major effect on the sensations received in the mouth. Ethanol acts as a partial anaesthetic, reducing the sensitivity of the palate towards acids and tannins in particular. When alcohol is removed from a wine, these substances produce an enhanced effect, and the wine tastes very acid and astringent. So, if you don't want to drink alcohol, drink tomato juice!

All alcohols have molecules that are constructed of chains of carbon atoms (C) with hydrogen atoms (H) attached, ending in the group that gives alcohols their special characteristics, the –OH group.   The different alcohols simply have different lengths of the carbon chain.

CH_3OH CH_3CH_2OH $CH_3CH_2CH_2OH$ $CH_3CH_2CH_2CH_2OH$

methanol   ethanol   propanol   butanol

The first in the series, with only one carbon atom is known as methyl alcohol, or methanol. This is a particularly unpleasant alcohol whose first effect is to make one feel very ill, followed by blindness, madness and then death. All wines contain some methanol, but in very tiny proportions. It is only a serious threat to health in badly conducted distillations: hence the tight control on illicit spirits in many parts of the world. Being a smaller molecule than ethanol, it boils at a lower temperature and in the production of spirits distils first as part of the 'heads' and can thus be discarded.

There are many other alcohols produced during fermentation, all of which have bigger molecules than ethanol. These are not as toxic as methanol but are still unpleasant in anything but trace quantities, causing headaches and nausea. As a group they are known as the higher alcohols or fusel oils. Because they have a higher boiling point than ethanol, they remain behind in the still at the end of distillation where they are a component of the 'tails'.

All of these alcohols have an important part to play in the maturation process of the wine. During this period they react with the natural acids to form the heady fruity substances known as esters. The familiar aromas of pineapples, bananas, strawberries, raspberries and most other fruits are due to the natural esters in the juice. Fresh grape juice does not contain many of these esters, but they are formed as a combined result of fermentation and maturation, hence the presence of the aromas of various fruits in the bouquet of a mature wine. Some of the higher alcohols have a powerful aromatic aroma of their own and contribute directly to the character of the wine. So it is not unreasonable, as some would suggest it is, to describe the characteristics of a wine in terms of other fruits and of flowers: the probability is that the same chemical compound is responsible in both cases.

Musts with high levels of sugar yield wines with a high concentration of alcohol unless steps are taken to halt the fermentation prematurely, which can happen naturally if the yeast is weakened initially by a very sweet must. High alcoholic strength is usually not the prime objective of any winemaker, as alcohol *per se* does not confer quality, although in hot climates it is difficult to avoid elevated levels of alcohol. In our present situation of climate change with rising temperatures (whether caused by human activity or not), grapes are getting riper and producing higher levels of sugar. This is resulting in higher levels of alcohol in the finished wine, which is not to the taste of the serious wine lover. In the middle of the twentieth century the norm for quality light wine was around 12.5% vol. Six decades later we are frequently seeing levels of 13.5%, with 14.5% being nothing unusual for wines from warmer climates. Even 15% can be found, although this is approaching the point at which extra tax would be levied in countries where the tax break is at 15%.

This higher level of alcohol has spawned methods of reducing the level by the use of physical processes of removing alcohol without harming the flavour of the wine. Reverse osmosis is one such method (see p.68), but the one that has caught the attention of the media is spinning cone technology. This somewhat mysterious machine is really nothing more than a fractional distillation process, where the wine is poured over a series of spinning cones that spread the wine into a thin layer so that the more volatile alcohol can evaporate away at a low temperature, while the rest of the components are unaffected. This caused controversy in the UK where wine that had been treated by this method in France was refused entry to the UK because spinning cone technology had not been approved by the UK authorities. This anomaly has now been corrected and such reduced alcohol wine is now happily available to those that want it.

It is, of course, possible to raise the alcoholic strength by a limited amount (controlled by regulations) by the addition of sugar prior to fermentation (see p.65).

Acids

Acids play a very important part of the constitution of wine. Without them, the colours would be strange, the flavour bland and the keeping qualities greatly reduced. Although tartaric and malic acids are the principle source of acidity, arising from the grape itself (see p.23), during fermentation many more acids are produced as a result of the biochemical action of the yeasts. Two acids always present as a direct result of alcoholic fermentation are lactic acid and succinic acid. A second source of lactic acid is the malo-lactic fermentation, where bacteria convert malic acid to lactic acid. Other acids variously produced are propionic, pyruvic, glycolic, fumaric, galacturonic, mucic, oxalic and others.

This complex array of acids has an important role in their reaction with alcohols, leading to the formation of esters (see below).

The chemical structure of the different acids is more varied than alcohols and is more difficult to comprehend, but the one thing they all have in common is the acid group –COOH that confers on the molecule the acid properties. The so-called total acidity of the wine is the sum of all the acid compounds, expressed as if it were all due to tartaric acid.

Control of total acidity is achieved principally at the must stage because experience shows that adjustments are best made before fermentation, since it would appear that the process of fermentation helps to blend the adjustments in some way (see p.57). However, it is permissible to make further adjustments of acidity in either direction after fermentation. In all cases, a laboratory analysis of total acidity is advisable in order to calculate the correct addition of acid (see p.235).

Esters

Esters are chemical compounds that result from the interaction of acids and alcohols. Many have pronounced fruity aromas, and indeed some are present in the grape and impart fruitiness to the grape juice. Many more are formed during the anaerobic maturation of the wine, but some are hydrolysed (split apart) by water into their component acids and alcohols, so a vast array of chemical reactions occurs during this period.

The principal ester is ethyl acetate which is the substance that results from volatile acidity, which is often incorrectly regarded as a derogatory term when in fact it is a descriptive one. Volatile acidity is that acidity which is due primarily to acetic acid, the acid of vinegar. All wines contain a certain amount of acetic acid, a natural component of their constitution, which adds to their complexity. It is only a fault when in excess, which can be prevented by good housekeeping.

Acetic acid is produced by the oxidation of alcohol, particularly under the influence of acetic bacteria (acetobacter).

$$CH_3CH_2OH \quad + \quad O_2 \quad \rightarrow \quad CH_3COOH \quad + \quad H_2O$$

$$\text{ethanol} \qquad\qquad \text{oxygen} \qquad\qquad \text{acetic acid} \qquad\qquad \text{water}$$

These bacteria are aerobic, needing oxygen to do their dirty work. Hence, quality control in this area is simple: the wine should be kept free from bacteria and away from oxygen. Unfortunately, grapes are swarming with acetobacter and the Earth's atmosphere contains 20% oxygen, so distinct action has to be taken to avoid trouble. A well-equipped modern winery has efficient filtration equipment for removing bacteria and a gas-blanketing installation (using nitrogen or carbon dioxide, or a mixture of both) for sweeping out air from vats and pipelines. Sulphur dioxide controls the acetobacter because they are very sensitive to it and are soon calmed down by a judicious dose at the right moment.

Wines suffering from excess VA do not usually smell of vinegar, but of nail varnish remover or cellulose thinners because some of the acetic acid forms ethyl acetate, which has a much stronger smell than acetic acid. The reaction going on is identical to that which helps to form the bouquet during maturation in bottle; it is the process known as esterification, when acids react with alcohols to form esters, which all have powerful volatile aromas.

$$CH_3CO\,OH \ + \ H\,OCH_2CH_3 \quad \rightarrow \quad CH_3COOCH_2CH_3 \ + \ H_2O$$

$$\text{acetic acid} \qquad\qquad \text{ethanol} \qquad\qquad\qquad \text{ethyl acetate} \qquad\qquad \text{water}$$

The EU limit for VA in red wine is 1.2 g/litre expressed as acetic acid, which, for most wines does not present a problem because VA becomes noticeable at levels around 0.8 g/litre. Difficulties in complying with the regulations occur only in older wines, where the VA can rise with time, and with wines from hot climates such as the eastern Mediterranean and North Africa, where the natural VA is higher due to the increased rate of activity of bacteria with the higher temperatures.

White and pink wines very rarely exhibit excess volatile acidity, principally because the grape skins, which are the prime source of the bacteria, are removed early in the process of winemaking. The EU limit for these wines is accordingly set at the lower level of 1.08 g/litre as acetic acid, although this seems to be very generous for wines that have 0.4 to 0.5 g/litre as a normal level.

Other esters contribute to the complex, fruity, heady aromas that typify a wine that has undergone its period of anaerobic maturation, probably in bottle. They are the esters of all of the various acids that have reacted with the whole range of alcohols, some of which have been split up by hydrolysis and have re-combined in different ways, resulting in that wonderful, fascinating aroma that typifies a fine wine, ready to drink.

Residual sugars

Residual sugars is the name given to the residue of sugars that is left when a fermentation has come to an end. The trio of alcohol, acidity and residual sugars forms the balance of all wines. At the natural end of fermentation, when most of the sugars have been consumed by the yeast, a dry wine will result with around two to three grams per litre of residual sugars. Many branded wines intended for everyday consumption are made to a standard specification by blending a dry wine with a measured quantity of unfermented grape juice. This can be either as *Süssreserve* (preserved grape juice) or rectified concentrated grape must (RCGM), which is grape juice that has been concentrated and purified. Ordinary cane or beet sugar, which is sucrose, is not allowed for the purpose of sweetening finished still wine; it is allowed only for enriching grape must before fermentation and for sweetening sparkling wine. By this method of adding a sweetener, it is possible to create dry, medium and sweet wines from the same base wine.

The residual sugar is mainly fructose, because most wine yeasts will ferment glucose preferentially, their cell membranes being more permeable to glucose. If a sparkling wine has been sweetened with sucrose, none of this will be present by the time the bottle reaches the table. All of the sucrose will have been changed by the acids in the wine to a mixture of equal parts of glucose and fructose, a process known as inversion. When a must has been enriched with sucrose before fermentation, the sucrose first undergoes this inversion to glucose and fructose, and is then finally converted to alcohol by the action of the yeast. Technically, it makes no difference whether a must or a wine has been sweetened by grape sugars or by sucrose; it is purely a matter of legislation.

The EU, with its propensity for regulation, has introduced legal definitions of the various descriptions of sweetness and dryness. One cannot help but question the purpose of such legislation, as there is no tax on sugar, and what really matters is the taste of the wine and not its analytical parameters.

The basic limits for sugar for each category are easy to understand, but the augmented limits for increased acidity levels are complex and somewhat bureaucratic, even if they do recognise the relationship between sugar and acidity.

- Dry wines – Up to 4 g/litre of sugar, or up to 9 g/litre where the level of total acidity expressed as tartaric acid does not fall more than 2 g/litre below the final residual sugar content.

- Medium Dry – Up to 12 g/litre, or up to 18 g/litre where the total acidity is not more than 10 g/litre less than the final residual sugar content.

- Medium or Medium Sweet – Exceeds the level for Medium Dry, but does not exceed 45 g/litre.

- Sweet – Not less than 45 g/litre.

By way of a footnote, residual sugars are sometimes referred to as reducing sugars. This is because most of the residual sugars are glucose and fructose, which have reducing properties and will react with an oxidising chemical known as Fehling's Solution (see p.239).

Glycerol

Glycerol, or glycerine, is a major by-product of fermentation and is frequently the next most abundant constituent of wine after water and alcohol. Originating from the sugar in the grape juice, it is not surprising that the higher the sugar content of the must, the higher the concentration of glycerol, although this can be affected by the yeast strain used. Thus, wines from hotter regions generally have higher concentrations than those from cool climates.

In its natural, pure state it is a colourless, viscous liquid with a slightly sweet taste. It plays a considerable role in the mouth feel of wine, imparting a smoothness and weight, without giving it overt sweetness. The 'legs' that appear round the edges of wine in a glass are thought to be due to the surface tension effect of a combination of glycerol and alcohol. It is not surprising, therefore, that heavy-weight wines from the hotter parts of the globe show the most pronounced legs in the glass.

Grapes that have been affected by *Botrytis cinerea* already contain glycerol as a result of metabolism of the grape sugars by the noble rot. The additional glycerol produced during the subsequent fermentation gives botrytised wines their supremely smooth viscosity, in which the total glycerol can reach levels as high as 30 grams per litre.

Aldehydes and ketones

Acetaldehyde is interesting in that it is a precursor of ethanol during the fermentation process, but is also a product of the oxidation of ethanol. When alcohols become oxidised, they produce aldehydes, ethanol producing acetaldehyde (ethanal). It is the most important aldehyde, constituting over 90% of the total aldehyde content of most wines.

In fino sherry it is an important constituent of the flavour profile, but in most other wines it is regarded as unwelcome, in that it confers on the wine a stale flavour. It also binds very strongly to sulphur dioxide producing bisulphite addition compounds and removing the protection of the free SO_2. Wines which have been badly made or badly handled exhibit this phenomenon, causing problems in keeping the free SO_2 at an acceptable level, and giving the wine a reduced shelf-life. Nowadays, with the great improvement in winemaking that has occurred over the last decades, this problem is rarely seen.

Ketones are also the products of oxidation, not of ethanol, but as part of the complex metabolism of sugars. The ketone that has the most pronounced effect in wine (and beer) is diacetyl, a substance with a powerful buttery or toasty aroma. Its production is particularly pronounced during the malo-lactic fermentation, hence the characteristic nuances created in wines that have undergone this treatment.

Unfortunately, ketones also have a powerful binding effect on sulphur dioxide, thus adding to the problem caused by acetaldehyde.

Chapter 13

CLARIFICATION & FINING

Wine is considered with good reason as the most healthful and the most hygienic of all beverages.

Louis Pasteur 1822-95

Is treatment necessary?

A new wine fresh from the fermentation is not a pleasant beverage; it is cloudy with yeast cells, has a dank nose of decomposition and can cause severe bowel problems if drunk in quantity. Yet beneath this dark blanket lies the embryo of a lovely, pleasurable and health-giving drink. All that is necessary is a gradual stripping away of the substances that are covering up the true nature of the wine. Much of this would occur naturally, given adequate time, but careful intervention, expertly applied, can shorten this period.

Wine is a complex mixture of natural substances, many of which are in a constant state of change. Despite modern analysis and the application of treatments both ancient and modern, wine in bottle sometimes contains solid matter, either in suspension or sitting on the bottom of the bottle. The deposits could be either proteins or tartrate crystals, both of which are natural wine components, or they could be the result of reactions between the various minerals in the wine, or between the proteins and the tannins. One thing of which we can be certain is that the deposits are harmless because harmful organisms cannot thrive in wine, owing to the presence of alcohol and acids.

Whereas knowledgeable wine drinkers happily accept, and even expect, deposits in bottles of fine wine, the vast majority of consumers who drink simple wine regularly expect the wine to be star bright to the last drop. In a way this is a pity because virtually every time wine is handled or treated a tiny portion of the quality is lost. The principle of good winemaking should be minimum treatment and minimum interference, a principle which is known as 'low intervention' winemaking. Nevertheless, to satisfy the requirements of the majority, there are treatments and additives available to reduce the probability of deposits. These treatments are intended to produce a stable product which will remain clear in the bottle and enable the consumer to drain it to the last drop.

It is important to distinguish between additives that remain in the wine until it is consumed and those that should be regarded as processing aids. The latter are added for the purpose of reacting with, and thereby removing, certain substances from the wine that would cause instability at a later date. During this process, the added substance is removed from the wine in combination with the natural constituent, and is therefore not present when the wine is consumed and should not be regarded as an ingredient. All of the substances discussed in this chapter are processing aids and should not be included in an ingredient list (whenever this becomes mandatory). Unfortunately, it is very difficult to prove that not a single molecule of such a substance remains in the wine at the end of the process, and it would be wrong to state that wines clarified with animal derivatives are suitable for vegetarians and vegans.

Racking

Once the bubbles of carbon dioxide have stopped rising, indicating that the fermentation has been completed, the first natural clarification takes place due simply to gravity. The dead yeast cells and cellular matter from the grape fall to the bottom of the fermentation vessel, forming what is known as 'the gross lees', meaning the initial large deposit. This deposit must not be left in contact with the wine because it will decompose, imparting to the wine an unpleasant, bitter taste, a condition known as 'yeast bitten'.

As soon as the wine has shown signs of becoming clear, which usually takes one to two days in a large tank, the wine is carefully drawn off, leaving the solid deposit in the bottom. This operation is known by the curious wine industry term of 'racking', which appears to be derived from the old English word 'rakken', meaning the skins, pips and stalks of grapes.

This process is too slow in a large modern winery, so use is made of the centrifuge to clarify the wine, in the same way that it would have been used to clarify the must prior to fermentation. The principle difference in use after fermentation is the danger of dissolving oxygen from the atmosphere, so it is essential that the interior of the centrifuge be flushed with nitrogen before use.

Protection from oxidation

When the fermentation has finished, the wine loses the protection of the carbon dioxide that has been evolving throughout the fermentation and becomes prone to oxidation, a condition from which it will suffer for the rest of its life. The good winemaker is well aware of this and will take steps to ensure that all the necessary controls are in place.

Additionally, one of the consequences of the production of carbon dioxide during the fermentation is that all of the free sulphur dioxide will have been lost. It will have been swept out of the wine by the action of the bubbles. It is necessary therefore to add a fresh dose of sulphur dioxide at the end of fermentation, before any damage can be done. (See chapter 4 for further information on oxygen.)

Blending

After the first racking, although the wine is in a somewhat raw condition, an experienced winemaker can get some idea of the quality of the finished wine. It is at this stage that the first round of tasting occurs, with a view to the ultimate blend that will make up the final wine.

To many people, both those in the wine trade and those who simply enjoy wine, blending is often regarded as a somewhat dubious operation, a means of getting rid of second-rate wines. This is a false image of what is a very important aspect of wine production. Blending ranks with fermentation as a critical part of the winemaking process. Virtually all wines produced anywhere in the world are blended before being bottled. The *'grand vin'* of a classed growth Bordeaux château, for example, will be blended from the best vines, the best parcels of the vineyard, the best barrels etc. Sometimes the addition of as little as five per cent of another permitted variety is added, which can have a major effect on the style of the wine.

The production of a medium price branded wine probably involves an incredibly complex blending operation, with components from many different sources:

- different grape varieties
- juice from different pressings
- fermentation at different temperatures
- fermentation with different yeasts
- fermentation in wood
- storage in stainless steel
- maturation in oak

By this means, the winemaker can control the style of the finished wine to a very close degree, and can even make several different wines from the stock of bulk wine at his disposal.

It is important that the blend is created before the clarification and stabilisation treatments take place because the blending of different wines can upset the balance that has been created by the various processes, rendering the wine unstable again.

Fining

Fining is an ancient process of clarification that is universally used for two purposes: the prevention of haze and the removal of some of the tannins to improve the balance of the wine. The active component of most of the fining agents is a naturally occurring protein, used nowadays in purified form, which modern winemakers find more reliable and easier to use.

The mechanism of fining is complex and is still not fully understood. It has long been known that wine contains a complex mixture of molecules and particles of many different sizes and types. For some reason, these particles are electrostatically charged, some negatively, some positively.

Freshly made wine contains three main groups of substances:

1. The simple molecules such as alcohols, acids and sugars. These substances form solutions in the wine, are a necessary part of its structure and remain in it through all the treatments.

2. The large particles such as pieces of grape cell, yeasts, and other substances, all of which contribute to the cloudiness of the wine. These substances can all be removed by the straightforward process of filtration.

3. Proteins whose molecules are large, but not sufficiently large to render themselves visible by causing cloudiness. They are also not large enough to be removed by a filter. These constitute the group of substances known as colloids.

Colloids themselves can be divided into two types, the stable and the unstable colloids. The stable colloids cause no problems, being invisible in solution and remaining so. In fact, one of the stable colloids, acacia (gum arabic), the same substance that used to be used for gumming paper and known as Gloy, is sometimes added to wine to increase its stability. It is the unstable colloids that have to be removed, because they will make the wine cloudy after bottling.

The molecules of these unstable colloids carry an electrostatic charge when youthful, the same charge for each molecule. Molecules with the same electrostatic charge repel each other, which keeps them apart and renders them invisible in solution. With age, when the protein denatures, the molecules rearrange themselves and lose the electrostatic charge, which enables them to clump together and form solid matter. Although totally harmless and tasteless, these colloids have to be removed if the wine is to remain clear and bright. Unfortunately their molecules are too small to be removed by filtration, so they have to be removed by the process known as fining.

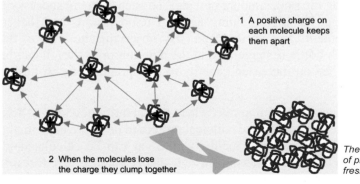

1 A positive charge on each molecule keeps them apart

2 When the molecules lose the charge they clump together

The natural state of proteins in freshly-made wine

The process of fining is very simple, which is why it is so widely used and has been for generations. In essence it achieves the same result as would occur naturally, given sufficient time. If another colloid with the opposite charge is added to the young wine, the oppositely charged molecules attract each other and form a solid precipitate which can be removed either by allowing it to settle or by filtration.

The removal of these colloids cannot be achieved by filtration alone because the colloidal molecules are too small to be retained by even the finest of filters. Fining and filtration are not interchangeable but are complementary: fining removes colloids, whereas filtration removes solid particles, including those formed as a result of fining.

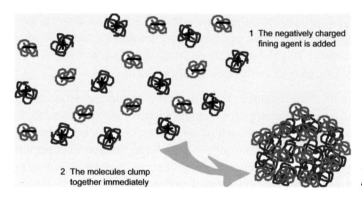

1 The negatively charged fining agent is added

2 The molecules clump together immediately

The action of a fining agent

The one problem which can occur is that if excess fining agent is added, the fining agent itself could precipitate with time, so it is important that no more than the correct amount is added. This can easily be determined by a simple experiment. A row of bottles is prepared, with each containing the same amount of wine. To each bottle in succession is added a larger quantity of fining agent. The bottles are then shaken to mix the contents and left to stand. By observing the level of the deposit in each bottle (exaggerated in the diagram opposite), it can be ascertained when an increased quantity of fining agent has no further effect.

This will be the correct fining rate, where the quantity of added colloid exactly matches the amount of colloidal protein to be removed from the wine. The correct quantity for the whole vat can be calculated by multiplication, the powder weighed out, dispersed in a few litres of

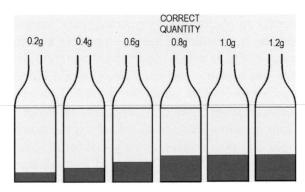

*The results of a fining trial
(greatly exaggerated)*

wine and added to the vat. After stirring thoroughly, the wine is left to stand until the deposit has fallen to the bottom, after which it is racked off into a clean vat.

Fining agents

Each fining agent has its own particular property and can be used for making a specific improvement in the wine. Many of the fining agents are themselves proteinaceous substances, obtained from natural sources: albumin from egg white, egg white itself, gelatine from bones and hides, casein from milk, isinglass from sturgeon.   All have molecules with the opposite electrostatic charge to the charge on the wine colloids, which they attract and remove.

Whilst some vegetarians will happily eat eggs or milk, vegans do not consume anything that is based on any product of animal origin. By the very nature of the fining process, the finished wine does not contain any of these substances, even if they have been used, because all of the fining agent will have been removed by reacting with the colloids in the wine.   Fining agents should be regarded as processing aids, not ingredients. The only sure solution to this problem is to use only bentonite – or no finings at all.

•   Ox blood

When asked to name a fining agent most people will quote 'ox blood'. However, this substance has *not* been permitted in Europe since the introduction of the winemaking regulations in 1987. It works, and used to be widely used. It is easy to use by simply pouring some animal blood into the vat and stirring it. The active agent is a protein known as albumin which is very effective at removing troublesome colloids. Winemakers continued to use albumin extracted from animal blood until 1997, when in response to concern about the spread of bovine spongiform encephalopathy (BSE or 'mad cow' disease), the French forbade the use of albumin prepared from cows' blood. It is known that the prion responsible for this disease is very stable and can survive heat treatment, but whether it could be carried through into wine is unproven.

- ## *Egg white*

This is one of the oldest fining agents, and is still widely used for fine red wines. It has a gentle action and also removes some of the harsh tannins. It is applied at the rate of 3 – 8 egg whites per barrique, by breaking the eggs into a bowl (and discarding the yolks), adding some wine, whipping together and adding the mixture to the barrel, which is then thoroughly agitated.

The active constituent is albumin, which is now available as a purified powder that is effective for both red and white wines. This is preferred nowadays since the scare regarding salmonella in chickens – a groundless preference, as viable salmonella cannot exist in a wine environment due to the presence of alcohol and acidity.

- ## *Albumin*

See above under blood and egg white.

- ## *Gelatine*

Gelatine is produced by boiling animal skins and bones, followed by treatment with acids, alkalis or enzymes. Its structure is slightly similar to albumin and it has similar properties, viz. it will combine with the harsh tannins in red wine and thereby remove them. The result is a softer wine which is also more stable.

For treating white wine, gelatine is often used in conjunction with silica sol (see p.152).

A vegetable-derived version of gelatine is now available for those who want no connection with an animal product.

• *Isinglass (ichthyocol or Col de poisson)*

The original source of isinglass was the swim-bladder of the sturgeon and other fish, but it is now often produced from fish waste from canneries. It is a pure form of gelatine and has a gentle action. It is used mostly for fining white wines, where it gives a good clarity. It also has a long history of use for the clarification of beer.

• *Casein*

Milk is the source of casein, which is yet another protein, and is useful for decolourising white wines. Some winemakers use skimmed milk rather than the pure substance.

• *Tannin*

Tannins are sometimes added to wine in combination with gelatine, being added after the gelatine. The tannins used are not the same type as would be found naturally in the wine. They are extracted from oak galls and are very bitter and astringent and, as such, they will precipitate first with the gelatine, bringing down the colloidal proteins in the wine.

• *Bentonite*

This is one of the most widely used fining agents, and it breaks the rule in that it is not a protein but a form of clay that is mined in various parts of the United States. It should not be confused with kieselguhr, which is a filtration aid and not a fining agent (see p.183).

Bentonite, or montmorillonite, is an alumino-silicate clay formed from volcanic ash, and whose small particles acquire a negative charge when dispersed in wine and are thus ideal for removing the positively charged protein molecules. The advantage of using bentonite is that there is no danger of over-fining, but set against this is the fact that it has strong powers of adsorption and can reduce the fruit of the wine in both aroma and flavour. Also, it forms a voluminous deposit from which it is difficult to recover the wine, so excessive use results in wasted wine. Despite its popularity, bentonite is not a substance that should be used carelessly.

• *Silica sol (Kieselsol)*

The active substance is silicon dioxide, which can be produced in both positive and negative colloidal forms. It is often used in conjunction with gelatine for removing other protective colloids from white wine. Like bentonite, it is a mineral and is not an animal protein.

• *Polyvinylpolypyrrolidone (PVPP)*

PVPP is quite unlike any other fining material in that it is a plastic material which has been milled into fine particles. Its particular property is the removal of phenolic components from white wine, especially those that are suffering from 'pinking' or 'browning' resulting from mild oxidation.

• *Activated charcoal*

Permitted for use in white wines only, where it can brighten a dull looking wine by removing colours. It must be used with great care, as it can remove flavours as well, and is used in the production of vodka where only the taste of pure ethanol is required. It is available in the form of filter sheets impregnated with charcoal.

FINING AGENT	SOURCE	WINE	REMOVES
Bentonite	Earth mining	All	Proteins
Egg whites, albumin	Eggs	Red	Tannin
Gelatine	Bones, hides	Red, white	Phenolics
Isinglass	Fish swim bladders	White	Tannin
Milk or casein	Milk	White	Colour, tannin
PVPP	Manufactured	White	Phenolics
Silica sol	Manufactured	White	Fining agents

• *Allergens*

The European Directive 2003/89/EC on the declaration of allergens in foodstuffs, including alcoholic beverages, came into operation in March 2005 and has caused confusion ever since. (See p.165 regarding sulphur dioxide.) On the one side, the regulatory authorities are demanding that the use of these substances be declared as ingredients on the label. Winemakers, conversely, argue that, although used as processing aids, these substances do not remain in the finished wine. It would, indeed, be incorrect to state their presence.

All proteinaceous fining agents are covered by this legislation and therefore, if used, will have to be declared on the label. It would be wrong to state that the wine contains an allergen if it does not, but how can it be proved? Analysis of proteins at this very low level is both difficult and expensive. Much effort has been put into such work, and isinglass has been given the 'green light' by the European authorities, probably as a result of work done by the brewers' research associations, who have shown that there are no residues left in beer after fining with isinglass. Similar research work has yet to be concluded by the wine sector to prove that a similar situation exists with the use of albumin and casein. If this cannot be achieved by the end of 2010, they will have to be declared on the label. The result of this unsatisfactory situation is that many winemakers have already ceased using these fining agents, and are using bentonite only, which is a pity, as they give a good result.

Blue fining

Blue fining is a misnomer as it has nothing to do with the removal of troublesome colloids, but is a chemical process for the removal of excess iron and copper in the wine by the addition of a solution of potassium ferrocyanide, as discovered by Herr Möslinger at the beginning of the twentieth century. The ferrocyanide reacts with any iron (Fe) and copper (Cu) that might be present in the wine, producing a deep blue deposit of ferric or cupric ferrocyanide. This simple process neatly removes excess iron and copper from the wine.

$$K_4Fe(CN)_6 \quad + \quad 2Cu^{2+} \quad \rightarrow \quad Cu_2Fe(CN)_6 \quad + \quad 4K^+$$
potassium ferrocyanide · · · · · · · · · · blue precipitate

Iron and copper are elements essential to all forms of life and are found in the soil and in all foodstuff, but in excess they cause problems, so it is important that the content of these two elements in wine should be controlled. First, both iron and copper can cause a haze or even a solid deposit in wine (see p.256).

Second, copper has a yet more fateful property in that it acts as a catalyst to oxidation, a catalyst being a substance that helps another reaction to take place. In the absence of copper and the oxidising enzymes the rate of oxidation of wine is quite slow.

There is a third and very basic reason for controlling the level of copper in a wine, namely that copper is a toxic metal at higher concentrations and has a legal limit of 1 mg per litre imposed by food regulations. Many wineries used to use bronze pumps and vat fittings, and some still do! These undoubtedly contribute to an increased level of copper in the wine because the acids in the wine readily attack the bronze, which is an alloy of copper and tin.

The process is harmless so long as a small residual quantity of iron remains in the wine, but residual ferrocyanide should be avoided because, although harmless in itself, it could be converted to cyanide, which is not exactly a pleasant substance! For this reason, blue fining can only be carried out when supervised by a qualified chemist.

Blue fining is particularly useful for removing excess copper, this being the metal that is removed first. The process can be used for all wines and is simple to apply, the only caveat being that a laboratory trial must be carried out first because the action of the ferrocyanide is somewhat unpredictable due to the possible presence of other metal ions which might interfere.

To ensure that no excess ferrocyanide is left in the wine, the quantity of potassium ferrocyanide added is such that one to two milligrams per litre of iron is left in the wine, which guarantees that no ferrocyanide remains in solution and the wine remains harmless. Suspicion is always aroused if the analysis of the wine shows zero iron, as all wine contains some natural iron.

The sequence of actions is as follows:

1. The wine is analysed to determine the level of iron and copper.
2. The theoretical quantity of potassium ferrocyanide is calculated.
3. A sample is treated in the laboratory to determine the effect of the theoretical addition. Adjustments to this are made, as necessary.

4. The full quantity of potassium ferrocyanide is calculated and weighed out.
5. After dissolving in water, the solution is added to the wine, which is then stirred.
6. After settling, the wine is racked off the deep blue sediment which contains the iron and copper.

Although blue fining is prohibited in many countries, in those where it is permitted, e.g. Germany, it is still used. Its advocates claim that it has no effect on the quality of the wine, but, as with so many wine treatments, there are those who prefer not to use it because they claim that it takes out fruit as well as the metals. The best control is undoubtedly prevention rather than cure, by eliminating the source of the contamination in the first instance.

Calcium phytate

Phytic acid is a naturally occurring substance found in the bran of cereals. It can be used in the form of its calcium salt to remove excess iron from red wines by precipitating it out as the very insoluble ferric phytate. A residue of phytic acid in the wine is undesirable because it combines with the essential calcium in the body, rendering it unavailable for metabolism. So the same principle applies as in blue fining, of leaving a trace of iron in the wine rather than residual calcium phytate. The maximum dose permitted is 8 g/hl. It would appear to be a rarely used treatment.

PVI/PVP copolymers

PVI/PVP is the acronym of polyvinylimidazole – polyvinylpyrrolidone, both of which are types of plastic material which have been polymerised together. It is used in fine granular form and has the property of absorbing metal ions such as iron and copper, and is an alternative to blue fining. There are strict rules regarding its use:

- It is allowed to be used on both musts and wines.
- The total used must not exceed 500 mg/l.
- It must be removed after no more than two days.
- The treatment must be under the control of an oenologist.

Chitin-glucan complex and chitosan

Chitin-glucan is of fungal origin and is a natural polymer, being the main component of the cellular walls of *Aspergillus niger*. This fungal resource is a by-product of citric acid produced for the food and pharmaceutical industries.

It was authorised in 2009 by the OIV as a fining agent for the clarification of wine because it will precipitate proteins in suspension.

Also authorised in 2009 by the OIV, chitosan is another natural polysaccharide produced from *Aspergillus niger* or *Aspergillus bisporus* and will remove heavy metals such as iron, lead and cadmium. It will also reduce the level of ochratoxin A and, amazingly, will eliminate undesirable organisms, notably *Brettanomyces*.

These substances appear to be used together as one treatment, but little information is available because they are new additions to the wine treatment armoury.

Chapter 14

TARTRATE STABILISATION

Every quotation contributes something to the stability or enlargement of the language.

A Dictionary of the English Language
Dr Samuel Johnson 1755

Natural and harmless?

The removal of unstable colloids and solid particles yields a clear, bright wine. Unfortunately, this does not mean that the wine is totally stable and will not deposit any further solid matter. The chances are high that, at a later date, it will deposit tartrate crystals in the bottle. Although these crystals are entirely natural and totally harmless, the average consumer will object to them and will return the bottle with a serious complaint. An attempt, therefore, has to be made to prevent this happening, but it is not easy to guarantee success and it is probably the biggest problem facing all wine producers. No supplier is prepared to give a warranty that their wine will never deposit tartrate crystals.

The mechanism of crystal formation is complex and involves the principle of "super-saturation". The naturally-occurring tartrates in the unfermented must are completely soluble because they are below the concentration at which their solution becomes saturated. However, after fermentation they are much less soluble because of the presence of alcohol; their concentration exceeds the solubility limit and they try to crystallise out because the solution is super-saturated. However, the colloids act as protective agents, preventing the formation of crystals. But this protective effect is only temporary. After a few weeks, or months, the colloids denature (change their properties), thus losing this protective effect and the tartrates can start to crystallise. This unfortunately happens in the bottle, causing complaints of 'broken glass' or 'sugar crystals' (see p. 252).

The two refrigeration processes described below depend upon the principle of the reduced solubility of a substance at lower temperatures. This works quite well for the removal of potassium bitartrate, which is the major cause of complaints about crystals. It does not work at all well for the removal of calcium tartrate because the solubility of this substance changes very little at low temperatures. Before commencing any of the tartrate stabilisation processes, it is essential that a thorough fining operation is carried out to minimise the level of the protective colloids.

Cold stabilisation

In the traditional process of cold stabilisation the wine is chilled to just above its freezing point, which is -4°C for a wine of 12% alcohol and as low as -8°C for a fortified wine, and is then stored in insulated tanks for up to eight days. This is expensive in terms of capital outlay for the refrigeration unit and a bank of insulated tanks, uses a large amount of energy for chilling and involves an expensive stock holding. What is more, the results are not always reliable!

Cold stabilisation installation in a bodega in Jerez

The process depends upon crystallisation being initiated by any minute particles that happen to be present that can act as nuclei. Once small crystals have formed, further crystallisation occurs at the surface of the crystals, which gradually grow bigger. The process is inefficient because the deposited crystals lie in the bottom of the tank, with convection currents being the only means by which the contents of the tank come into contact with the crystals.

The weakness of this process is that colloids and other components in the wine have a very big influence on the efficiency of the crystallisation process. They have a protective effect on the micro-crystals, somehow preventing further deposition. The result is that, at the end of the cold treatment, the concentration of potassium bitartrate in solution has not been reduced sufficiently to guarantee stability after bottling. Later in the life of the wine, when the colloids have ceased to have their protective effect, the crystals form inside the bottle. Hence the critical importance of an efficient fining process before refrigeration.

Contact process

This weakness in the old static process led to the development of the contact process, which has proved to be quicker, cheaper and more effective. It gets its name from the principle of bringing the wine into contact with micro-crystals of potassium bitartrate, which act as nuclei,

enabling the crystallisation to occur more readily. The process starts with the chilling of the wine, but not to the low temperatures required by the traditional process: 0°C is usually sufficient. Finely ground crystals of potassium bitartrate are added at a rate of 4 g/l and the wine stirred vigorously for one to two hours. The vigorous stirring keeps the crystals in suspension, thus ensuring close contact with the wine so that they can function by drawing out the excess tartrate, growing bigger in the process. At the end of the stabilisation the wine is filtered while still cold, and the crystals are separated out, ground down and added to the

Crystal separator at a famous sherry producer next batch of wine.

The latest development is the continuous contact process, where the crystals are packed into the conical base of a vertical tank and the cooled wine is pumped upwards through the crystal bed. The stabilised wine can be drawn off at the top of the tank. The tank has to be opened periodically and the enlarged bed of crystals removed.

Ion exchange

The process of ion exchange has been used in the domestic situation for many years in the form of the water softener, which consists of a container packed with an ion exchange resin. This resin is a form of plastic which contains loosely bonded sodium ions. Hard water contains calcium and magnesium salts which are the cause of the hardness. When the water flows past the particles of resin, the calcium and magnesium in the water are attracted to the resin, being replaced by the sodium ions from the resin. When the resin is saturated with calcium and magnesium, brine, which is a solution of common salt or sodium chloride, is passed through the resin bed. Now the reverse happens: the sodium goes on to the resin and the calcium and magnesium come off, and are flushed down the drain.

Ion exchange
equipment
as used in
Australia

A similar process occurs in the treatment of wine: the potassium and calcium in the wine are attracted to the resin and are replaced by the sodium ions. The wine now contains sodium bitartrate. This prevents the formation of tartrate crystals because sodium bitartrate is much more soluble than the calcium and magnesium salts. Thus, ion exchange is the prevention of crystallisation, whereas refrigeration methods rely on the removal of excess tartrates.

Unfortunately, from a health aspect, it is not a good thing to replace potassium with sodium. It is common knowledge that excessive salt is bad for the body, especially in relation to cardio-vascular conditions. We are all told not to put too much salt on our food. One of the simple counter measures is to administer a potassium salt, such as potassium chloride, which replaces the sodium in the body and rectifies the situation. Wine is one of the richest sources of dietary potassium and must be a much more pleasurable way of taking potassium than swallowing potassium chloride tablets! Here we have yet another of the benefits of drinking wine, and not merely for pleasure. Ion exchange removes this health-giving element and replaces it with an unwelcome entity, so it is not surprising that the use of ion exchange was prohibited within the European Union. However, it would appear from Regulation (EC) No 606/2009 Annex 1A #43 which states 'cation exchange can be used to ensure the tartaric stability of wine' that it is now allowed in Europe.

Electrodialysis

This is a comparatively new technique that uses the property of special selective membranes to allow the passage of potassium, calcium and tartrate ions under the influence of an electric charge.

The membranes can be tailored to suit a specific task and the operation of the machine can be monitored by constant measurement of the conductivity of the wine being treated. The main disadvantage is the high capital cost, so it is really only suitable for large wineries.

The advantages over any form of refrigeration are :

* Much lower energy costs.
* Treatment is tailored to each wine.
* Both potassium and calcium tartrates are removed.
* Wine does not need a lot of pre-treatment.
* Results are reliable.

Electrodialysis
equipment
at a large
bottling plant
in Bordeaux

Metatartaric acid

Metatartaric acid is a strange substance of no defined structure which is
produced by heating tartaric acid to a high temperature in a closed
vessel. During this process the molecules partially polymerise, or bind
together. It dissolves readily in water or wine and has the property of
preventing the deposition of tartrate crystals, but the precise mechanism
is uncertain. It is thought that it coats any micro-crystals that might be
present in the wine and prevents their development to a visible size. In
view of the uncertainty of some tartrate stabilisation processes, this is a
useful substance, and is cheap and effective.

This might seem to be the perfect answer to the prevention of tartrate
crystals in the bottle, but unfortunately metatartaric acid is unstable in
solution and gradually reverts to ordinary tartaric acid, resulting in the
possibility of even more tartrate crystals. This happens faster under
warmer conditions, as might be expected, in the knowledge that all
chemical reactions proceed faster at higher temperatures. The effective
life of metatartaric acid in wine at 25°C is about six months, whereas at
10°C it will last for eighteen months. It is therefore ideal for wines
packed in bag-in-box, which should be consumed within twelve
months. The formation of tartrate crystals in this type of packaging is

particularly serious, leading to leaking taps and to complaints of ruined furniture and carpets. It is of no use for the protection of wine intended for long periods of maturation in bottle.

It is also useful in those circumstances where tartrate stabilisation has not been possible, as in small wineries without sufficient capital for the installation of expensive refrigeration equipment. Indeed, there are large modern wineries that prefer to use metatartaric acid rather than cold stabilisation which, they feel, could adversely affect the quality of the wine.

The limit imposed by EU regulations is 100 mg/litre, this being the level normally used in practice, there being little point in using less than the maximum dose.

Carboxymethylcellulose (Cellulose gums)

Somewhat like metatartaric acid, cellulose gums prevent the deposition of tartrate crystals by interfering with the formation of crystals by coating any potential sites of crystallisation. The advantage is that the effect lasts longer than metatartaric acid (see above). The limit imposed by EU regulations is 100 mg/l.

Mannoproteins

It has been noted for some time that wines that mature on the lees have greater tartrate stability. As a result of work carried out on model wine solutions, it was discovered that this is caused by the mannoproteins released during yeast autolysis. These act as a protective colloid which covers the surface of the crystal nucleus used for the crystallisation and the crystallisation process is therefore unable to proceed. These mannoproteins can be produced on a commercial scale by enzymatic hydrolysis of cells of *Saccharomyces cerevisiae.* The purified mannoprotein is supplied as a white powder that is soluble and has no colour, flavour or taste. It is permitted for use in the European Union and Argentina and is currently being investigated for use in Australia and the USA . It must be added to the wine the day before bottling at a dosage of 200 to 250 mg/l and good mixing is obviously important. Unlike metatartaric acid, the effect is long-lasting because the protein molecules are relatively stable.

Minimum intervention

Although none of the above treatments, if properly carried out, has a noticeably damaging effect on the quality of the wine, the best philosophy is to use the minimum treatment possible. The principle of 'low intervention' winemaking, as this is known, is undoubtedly the correct approach. However small the effect might be, in every stage of handling there is the potential for a loss of quality and an increase in costs.

If we could but convince the general wine drinking public that deposit in a bottle of wine indicates that the wine has received minimum treatment and is in a more natural condition, we would all be winners.

Chapter 15

ADDITIVES

The best use of bad wine is to drive away poor relations.

French proverb

The additives allowed for wine, in common with those allowed for any foodstuff, are strictly limited by regulations which vary from country to country. Well-made wine is a fairly stable product and should not need additives other than those to combat oxidation and the attack of microorganisms, and even these may not be essential. From a food safety angle, it is comforting to know that virtually no pathogen can survive in wine because of the presence of alcohol and the natural acidity.

All of the substances discussed in this chapter are true additives in that they remain in the wine until the point of consumption. They are therefore correctly described as additives, and should be included in an ingredient list, if such a list is mandatory. Ingredient listing for wine has been tossed around in Europe for many years, partly because it is difficult to decide whether an additive is an ingredient or whether it is merely a processing aid. Also, many wines, especially table wines, are highly blended products, blends that might vary from bottling to bottling and which would require constant changes to the ingredient list. Wine is still exempt from such a requirement at present (2010), and indeed it is illegal to put an ingredient list on a wine label as it is not one of the optional labelling items.

Sulphur dioxide

This most useful of all additives has been known for centuries, long before its action was fully understood. Sulphur is an element, with the chemical symbol S, which is found in the earth's crust in volcanic regions and is a pale yellow brittle rock-like substance. Surprisingly, when ignited it burns with a blue flame, the yellow solid melting to a viscous orange liquid. However, more is happening as becomes apparent when a lung-full of a noxious pungent gas is inhaled, causing severe breathing problems resembling an attack of asthma.

The gas is sulphur dioxide and is formed according to the following equation:

$$S \quad + \quad O_2 \quad \rightarrow \quad SO_2$$
sulphur $\qquad$ oxygen $\quad$ sulphur dioxide

Sulphur dioxide gas is soluble in water, actually reacting with it to produce sulphurous acid (but see p.173):

$$SO_2 \quad + \quad H_2O \quad \rightarrow \quad H_2SO_3$$
sulphur dioxide $\qquad$ water $\qquad$ sulphurous acid

This is the sequence of events that caused the so-called "acid rain" that was being produced in England before the power stations cleaned up their flue gases. Residual sulphur in the fuel was burned in the furnaces, producing sulphur dioxide which was released into the atmosphere. It was alleged that this made its way to Scandinavia where it dissolved in the rain, ultimately producing sulphuric acid which destroyed conifer forests. Irrespective of the scientific facts behind this allegation, the power stations are installing expensive equipment to remove the sulphur dioxide before the flue gases are released from the top of the chimney.

It can be seen that the colloquial use by winemakers of the word 'sulphuring' is misleading because it is not sulphur that is being added but sulphur dioxide. This habit has undoubtedly grown up due to the fact that, long before the chemical industry was born, sulphur dioxide was produced on the spot by burning a piece of sulphur. Empty casks were (and still are) sterilised after rinsing by lowering a piece of burning sulphur into the cask. The sulphur dioxide produced by this reaction dissolves in the residual water in the cask, thus sterilising it.

Although sulphur candles are still used, a convenient way of obtaining sulphur dioxide is by using potassium metabisulphite, a white powder with the chemical formula $K_2S_2O_5$. This substance has the useful property of releasing sulphur dioxide when dissolved in an acid aqueous liquid, which is precisely the constitution of grape must and wine. Under these conditions it releases 57% of its weight as sulphur dioxide. Alternatively, pure sulphur dioxide can be purchased in cylinders as the liquefied gas, from whence it can be metered into the wine (and this process can even be automated).   This method is particularly useful for direct addition to wine, using specially designed dosing equipment which can be adjusted to give exactly the right concentration.

Solid sulphur pellets for sterilising barrels   *Burning sulphur produces sulphur dioxide*

Although toxic in large doses, sulphur dioxide is harmless when used at the correct level. It is used in many different foodstuffs as an antioxidant and a preservative. It can be found in dried fruit, fruit juices and squashes, fresh fruit salads, sausages, peeled potatoes and many more foodstuffs. The World Health Organisation has conducted a study on the total sulphur dioxide in the diet and has concluded that the present levels are within safe limits. In Europe the permitted additives for wine are:

> E220   sulphur dioxide
> E224   potassium metabisulphite
> E228   potassium bisulphite.

It should be noted that sodium metabisulphite and sodium bisulphite are not permitted in wine, although they are widely in use for preserving fruit squashes and other beverages. The importance of observing safe limits is emphasised by the fact that people have died after eating fresh fruit salad which has been carelessly dosed with sulphur dioxide.

Although very useful, sulphur dioxide does have a disadvantage in that it can cause an allergic reaction in some consumers who are prone to asthma or other allergies. Such people have to be aware that most wines contain this substance, hence the labelling regulation that has been in force in the USA for some years that demands the statement, 'This wine contains sulfites'. And from 25 November 2005 all wines produced in the EU that contain more than 10 mg/litre of sulphur dioxide must be labelled 'Contains sulphites' or 'Contains sulphur dioxide' (see p.287). In effect this means that all wines have to be labelled thus because they all contain natural sulphites above the stated limit, being produced by yeasts during fermentation from naturally occurring sulphur compounds.

A second disadvantage is that sulphur dioxide bleaches the colour of red wine and can also result in a loss of fruit, although the wine does partially recover with time when the level of free sulphur dioxide has diminished naturally. The good winemaker always uses sulphur dioxide sparingly.

Sulphur dioxide has become such a universal additive because it has four quite distinct properties:

1 Antioxidant

The prime use of sulphur dioxide is the prevention of oxidation, the antioxidant property. In wine production it is now realised that oxidation must be kept to the minimum if fruit is to be conserved.

The reason for the anti-oxidant property of sulphur dioxide is that it will readily combine with oxygen, thus removing it before too much harm can be done. The product of this reaction is that familiar schooldays chemical, sulphuric acid, which might seem a totally inappropriate substance to find in wine, but the concentration is very low, amounting to a few parts per million and is harmless at this level. Only at much higher concentration does it become aggressive and dangerous.

$$H_2SO_3 \quad + \quad [O] \quad \rightarrow \quad H_2SO_4$$

sulphurous acid   oxygen   sulphuric acid

[Certain aspects of some of the equations shown are not strictly correct, in chemical terms, but have been simplified to show the principles of the function of sulphur dioxide (see p.173).]

Although sulphur dioxide will scavenge any oxygen that becomes dissolved in the wine, the reaction is not immediate. It is possible for oxygen and sulphur dioxide to exist together in the wine, before reaction takes place, and herein lies the danger. It is during this time that oxygen can damage the wine, so it is far better to ensure that oxygen and wine never come into contact by careful handling and by the total exclusion of air. Sulphur dioxide will not compensate for poor wine handling techniques.

The level of sulphur dioxide has to be checked many times during the processing and handling operations because its action is self-destructive. Every atom of oxygen destroys a molecule of sulphur dioxide, resulting in a constantly falling concentration. A particularly important moment is in the preparation of wine for bottling, for it is critical for the keeping qualities of the bottled wine (particularly wines for everyday drinking) that the sulphur dioxide is adjusted to the correct level.

2 Antiseptic (anti-microbial)

Septic wounds are caused by bacteria infecting the surrounding tissue, where they can flourish in the warmth and in the damp nutritious surroundings. When an antiseptic is applied to a wound the bacteria are killed. Sulphur dioxide has the same effect on bacteria in wine: they are easily killed, which is most fortuitous since acetobacter is probably the commonest bacterium to attack wine, turning it to vinegar.

Acetobacter are aerobic bacteria and need oxygen to flourish. A dose of sulphur dioxide followed by filtration is the immediate treatment for any wine that has such an infection. The sulphur dioxide will initiate a two-pronged attack by poisoning the bacteria and by removing any remaining oxygen that the bacteria need if they are to thrive.

The antiseptic property of sulphur dioxide is also used to prevent the malo-lactic fermentation. In this instance it attacks the lactobacillus, a bacterium which converts malic acid to lactic acid (see p.84).

Yeasts are more tolerant of sulphur dioxide but this tolerance varies from strain to strain, which provides a useful way of selectively subduing the less desirable ones that, fortuitously, tend to be less tolerant. The addition of sulphur dioxide before fermentation reduces the activity of wild yeasts and permits the wine yeasts to take over at an earlier stage, which gives a better and more secure fermentation.

It is a common misconception that sulphur dioxide at bottling is used to prevent a re-fermentation. Wine yeasts can tolerate a far higher level of sulphur dioxide than would be desirable, either for reasons of taste or to keep within the legal limits for total sulphur dioxide. Some microorganisms might have their reproductive rate slowed down, but the principal reason for using sulphur dioxide at bottling is for its anti-oxidative property.

3 Anti-oxidasic

The third property of sulphur dioxide relates to enzymes. The oxidation of fruits and fruit products is a slow process in the absence of any catalyst. However, we know from the observation of fruits such as apples that they brown very rapidly when they are cut or bitten, indicating that something must be present that can hasten this reaction. Indeed, there is something present acting as a catalyst, and that catalyst is an oxidase.

Sulphur dioxide acts as a poison to the oxidases (of which there are many different varieties), greatly reducing the rate of oxidation, and adding further to its antioxidant property. In fact, sulphur dioxide is used commercially to preserve things like fresh fruit salad, where it prevents the browning of white-fleshed fruits such as apple and pears. When added to wine it has the same effect and prevents the browning of white wines. (See p.178 for more on enzymes.)

4 Corrective after oxidation

The first three properties of sulphur dioxide are all preventative, but there is a fourth property that is corrective in that it can freshen tired wines which are suffering from a slight degree of oxidation. Wines in such a condition have probably been badly handled and have lost their entire free sulphur dioxide. The first action when dealing with such a wine is to analyse it and to make a suitable addition of sulphur dioxide to restore it to its correct level. However, this does more than simply protect it against further oxidation, it actually improves the taste of the wine by refreshing it.

The reason for this is that one of the main products of the oxidation of alcohol is acetaldehyde, which is a predominant component of the characteristic sherry nose. When added to a tired wine, sulphur dioxide combines with acetaldehyde, converting it into an odourless and tasteless compound, thus removing any hint of oxidation and restoring the wine (almost) to its youthful nature.

$$CH_3CH_2OH \quad + \quad [O] \quad \rightarrow \quad CH_3CHO \quad + \quad H_2O$$

ethanol   oxygen   acetaldehyde   water

$$CH_3CHO \quad + \quad SO_2 \quad \rightarrow \quad \text{tasteless bisulphite addition compound}$$

It is important to realise that sulphur dioxide is not the 'magic potion' that will correct all bad wines and bad winemaking. Good technique is necessary at all times. Sulphur dioxide merely helps to maintain quality: it cannot create it where it is absent.

• *Free and total sulphur dioxide*

Sulphur dioxide in wine cannot be discussed without reference to 'the free and total', an expression commonly used by winemakers. To understand the significance of free sulphur dioxide and total sulphur dioxide it is necessary to understand a little chemistry.

Sulphur dioxide is a reactive substance and will combine not only with oxygen but also with other natural substances in the wine, such as sugars, aldehydes and ketones. When combined with these substances, the sulphur dioxide no longer possesses any of its protective properties and is known as the combined, or bound, or fixed sulphur dioxide. Badly made wines and oxidised wines have a higher proportion of aldehydes and ketones than wine in good condition and this results in a greater proportion of the sulphur dioxide becoming combined. The more that is combined, the less is available for its prime purpose of protecting the wine.

$$SO_2 + \text{aldehydes or ketones} \quad \rightarrow \quad \text{bisulphite addition compounds}$$

That portion of the sulphur dioxide that is not combined is known, logically, as the free sulphur dioxide. This is the active substance that has the protective properties.

The sum of the free sulphur dioxide and the bound sulphur dioxide makes up the total sulphur dioxide, which is also logical.

$$Free\ SO_2\ +\ Bound\ SO_2\ =\ Total\ SO_2$$

The total sulphur dioxide is regulated by EU law. In the stomach much of the bound sulphur dioxide is released by the acid and the warmth of the stomach contents and so is free to do its damage to the body if present above the safe limit. The basic legal limit for wine is 150 mg per litre, which is the level applied to dry red wine, that is a red wine containing not more than 4 grams per litre of residual sugars. It was recognised that white wine needs more protection than red wine because the latter contains its own natural anti-oxidants, the polyphenols. So white wine is allowed an extra 50 mg per litre. Further, because of the binding power of sugars, another 50 mg per litre is allowed for wine containing not less than 5 mg per litre or residual sugars.

Still greater quantities of sulphur dioxide are allowed for the natural sweet wines of the world, which has resulted in a plethora of different legal limits for the many different wine styles.

EU LIMITS FOR TOTAL SULPHUR DIOXIDE

- Dry red wine — 150 mg/l
- Dry white wine and rosé — 200
- Red wine with 5 g sugar/litre or more — 200
- White wine with 5g sugar/litre or more — 250
- Spätlese, white Bordeaux Supérieur etc — 300
- Auslese etc — 350
- Trockenbeerenauslese, Beerenauslese, Ausbruch, Sauternes, Bonnezeaux, Graves Supérieures etc — 400

Years ago, in the days of bad winemaking, wines contained high levels of binding substances produced by poor techniques. It was sometimes a problem to maintain sufficient free sulphur dioxide without exceeding the legal limit for the total (which was higher then, in any case). Thankfully, the situation has improved greatly and this is no longer a problem. The progressive reduction in the permitted limit of total sulphur dioxide was undoubtedly one of the factors that forced winemakers in many parts of Europe to improve their techniques.

• *Molecular sulphur dioxide*

There is yet more to the sulphur dioxide story because the free sulphur dioxide exists in more than one form. The chemistry involved is quite complicated, but the details of this are not important provided that the concept of molecules splitting into positive and negative ions can be grasped.

When dissolved in wine or water, a chemical reaction takes place between the sulphur dioxide and the water forming sulphurous acid. This acid actually exists in solution not as H_2SO_3 but in three different ionised forms. At the pH of wine, the predominant form is HSO_3^-

$$SO_2 \quad + \quad H_2O \quad \leftrightarrows \quad H^+ \quad + \quad HSO_3^-$$

This reaction can go in either direction, indicated by the two arrows in the middle of the equation. Thus the two ions can re-combine to form water and sulphur dioxide. This type of reaction is known as an equilibrium because all four entities are present at the same time. The wine will thus contain both the ionised form of sulphur dioxide and the simple unionised molecule that is known logically as the molecular sulphur dioxide. It is only the molecular form that possesses the protective properties and not the ionised form. The molecular sulphur dioxide performs the important operations of scavenging the oxygen, killing the micro-organisms, poisoning the enzymes and freshening the wine.

The balance of this equilibrium is affected by the pH of the liquid. At lower pH levels, the reaction moves to the left, and at higher pH levels it moves to the right. This means that the proportion of molecular sulphur dioxide increases as the pH of the wine gets lower, meaning a more acid wine. (See p.237 for more information on pH.) Therefore acidic wines require lower doses of sulphur dioxide than those wines with higher pH and less acidity. The truly scientific winemaker will take into account the pH of the wine before deciding on the amount of sulphur dioxide to add, thus maintaining the molecular sulphur dioxide at the correct level.

Ascorbic acid

Ascorbic acid is the chemical name for vitamin C, the valuable vitamin found in many fresh fruits and vegetables. It has powerful antioxidant properties which make it a useful additive for anaerobic winemaking. It is for the same reason that it is of value in our diet, protecting our bodies against the effects of oxygen and helping to delay the ageing process.

The danger of using ascorbic acid as an anti-oxidant is that it produces hydrogen peroxide when it becomes oxidised, and hydrogen peroxide is a powerful oxidising agent. To make matters worse, the product of the oxidation of ascorbic acid is dark brown in colour, thus rendering the wine even darker than it would otherwise have been! Therefore the use of ascorbic acid can be total disaster, with the result being far worse than if it had not been used at all. It cannot be regarded as a substitute for sulphur dioxide, but rather as reinforcement.

The way in which this effect can be prevented is to ensure that sulphur dioxide is present as a protective agent for the ascorbic acid, and under these conditions the ascorbic acid does give extra protection. There are many trained winemakers who use it regularly, especially in white wine production in countries such as Germany, New Zealand and Australia, because, properly used, it keeps the wine ultra-fresh.

Ascorbic acid has what is known as an isomeric form, erythorbic acid, which is used in Australia instead of ascorbic acid because it is cheaper, yet has similar properties. (An isomer is a substance whose molecules contain the same number and types of atoms but in a different arrangement.) Erythorbic acid, however, is not permitted for use in wines sold in the EU, and this includes wines from Australia.

Being vitamin C, ascorbic acid is beneficial and is not harmful in any way, which makes the EU limit for wine of 150 mg/litre seem slightly odd. The human body can cope with any amount of vitamin C, merely secreting the excess in the urine. It was Linus Pauling, the American chemist (1901 – 1994), who suggested that taking a teaspoonful of ascorbic acid every day would protect against the common cold. The only adverse reaction is possibly an attack of diarrhoea!

Sorbic acid

Sorbic acid has one property that is useful in winemaking: it stops yeast fermenting and is used as an additive before bottling to prevent re-fermentation in bottle. It does not kill yeasts and therefore is not a fungicide, but merely prevents fermentation by interfering with the metabolism of the yeast. As it does not kill the yeast, reproduction can still carry on, ultimately producing a flocculent deposit which would be a just cause for complaint. The more serious conditions of cloudy wine and popping corks are, however, prevented.

This property of sorbic acid is dependent on the combined presence of sulphur dioxide, alcohol and acidity, which is convenient for wine producers because all three of these conditions are satisfied. However, the efficacy drops with lower alcohol levels, demanding higher dosage rates. The EU limit is 200 mg/litre, at which level some people can begin to notice the taste of sorbic acid (and some can taste it at much lower levels), so most wine bottlers add 150 mg/litre to all wine as a compromise. At this concentration, an alcoholic strength of about 12% is required if fermentation is to be prevented. In a wine of 10.5% alcohol, this level has little protective effect, and yet susceptible wines with residual sugar, such as German wines, have habitually been bottled with the addition of sorbic acid at this level, which would seem to be pointless.

Despite the reference to the use of sorbic acid, the actual chemical added is potassium sorbate, which dissolves readily in wine and is decomposed by the acids in the wine to become sorbic acid. The solid form of the pure acid dissolves only with difficulty.

It is important that the addition of sorbic acid is made just prior to aseptic bottling because it has no bactericidal properties whatsoever, and can be the source of an all-pervading smell of geranium leaves (strictly, pelargoniums) when metabolised by certain strains of bacteria. For those who really must know, this smell is caused by the formation of 2-ethoxycarbonyl-3,5-hexadiene. Wine in this condition is fit only for destruction, so it is important to ensure that wine containing sorbic acid is free from bacteria.

It is not surprising that its use is diminishing, with many retailers forbidding it, not for any health implications, but merely to reduce the use of additives. It should not be necessary, because careful filtration and good hygiene are all that is required for successful aseptic bottling. The need for sorbic acid has to be an admission of a lack of confidence in the bottling process.

Metatartaric acid

Metatartaric acid has the surprising property of preventing tartrate crystals from depositing in the bottle, but only for a limited time. Although a well-known property, the mechanism is uncertain. (For full details see p.162.)

Citric acid

Citric acid is the natural acid of citrus fruits and is totally harmless, as would be expected. Its use in winemaking is for treating wines with a high level of iron, where it is not possible to carry out a blue fining operation (see p.153). In these circumstances it has the useful property of preventing iron casse by forming a soluble complex with iron, thus preventing it from forming an insoluble compound with the natural phosphates in the wine. This is a simpler option than blue fining, although in this case the iron is not removed but merely complexed.

Although used in some parts of the world for acidification of wine, citric acid is not a natural component of grape juice, and is not allowed within the EU for that purpose, hence the EU limit of 1 g/litre. It is always added to the finished wine and never to the unfermented juice, because it can be converted into acetic acid by the action of the yeast, resulting in a wine with excess volatile acidity.

Copper sulphate or silver chloride

In these days of stainless steel equipment, it is ironic that there is a higher incidence of reductive taint (dirty drains or bad eggs) than with old-fashioned machinery. This is due to the fact that traces of copper from old bronze equipment, such as pumps and hose couplings, removes the hydrogen sulphide which is the source of the taint. The use of ultra-anaerobic techniques, where oxygen is never allowed to touch the wine, encourages the formation of hydrogen sulphide by the reduction of sulphur dioxide. Also, many yeasts produce traces of hydrogen sulphide from sulphur compounds during the fermentation. The complete absence of bronze components means that there are no copper ions to act as cleansing agents.

Under these circumstances the treatment is the addition of copper in the form of copper sulphate, an attractive bright blue crystalline powder well-known from the days of school chemistry. The quantity added must be carefully calculated after analysis, otherwise there will be too much residual copper in the wine, necessitating a blue fining operation. The copper content of the wine after the treatment must not be more than 1 mg/litre.

An informal demonstration of the effectiveness of copper can sometimes be useful in the tasting room. When a wine has a nose smelling of dirty drains or bad eggs, a bronze coin (e.g. a British 1 or 2p coin, or 1 or 2 € cents) should be added to the glass and the wine swirled for a minute or so. If the smell disappears and the wine assumes a healthy fruity nose, hydrogen sulphide is the cause; the copper dissolved from the coin by the acids in the wine has reacted with the hydrogen sulphide and precipitated it as copper sulphide. If there is no improvement, the cause of the taint has some other origin.

Silver chloride was added to the EU list of permitted treatments in 2009 and is used for the same purpose as copper sulphate.

Acacia (Gum arabic)

Acacia, gum acacia, or gum arabic, is a substance obtained from *Acacia senegal*, a shrub native to the Sudan and is probably best known for its use as a paper glue. Chemically it is a polysaccharide and is related to the polysaccharides found in grapes. (Polysaccharides are large molecules made up of simple sugar molecules.) It is classed as a stable colloid and will stabilise the unstable colloids (see p.147). Its use in wine is very ancient and it is added to young wine intended for early consumption to slow the precipitation of colouring matter.

It is important that the addition of acacia is made after cold stabilisation. It is a stable colloid and will prevent the crystallisation of tartrates during the chilling process.

Enzymes

Enzymes are most useful entities, but are not easy to define. They are not living organisms because they cannot reproduce on their own. It is therefore incorrect to state that they can be killed, but they can be poisoned.

All enzymes are catalysts, and all life depends on enzymes. If they are destroyed, life ceases. The reason that cyanide is such a deadly and rapid poison is that it poisons the enzymes in our bodies.

But what is a catalyst? A catalyst is a substance that enables a chemical reaction to proceed, but does not itself take part in the reaction. Enzymes increase the speed of chemical reactions.

They are named by words ending in -ase; thus, the enzymes that promote oxidation are known as oxidases.

Enzymes are used at various stages in the winemaking process, with the first addition sometimes being made to the grapes in the press. Others are added at the must stage or later.

• *Pectinolytic enzymes*

Grapes contain various pectins and gums that increase the viscosity of the juice, making it more difficult to separate from the structure of the grape and increasing the time taken for the solids to settle.

These substances are composed of complex molecules containing long branched chains of atoms which help to hold the structure of cells together. Enzyme preparations known as pectinolytic enzymes have been extracted from certain species of moulds and have the useful property of breaking down the pectin chains into smaller units. When added to the grapes in the press the viscosity of the juice is decreased and a better extraction takes place.

However, many winemakers have discovered that this treatment can reduce the varietal character of the juice, resulting in a bland product. But good use can be made of these enzymes if added to the must after pressing. At this stage the reduction in viscosity enables a quicker and more effective clarification to take place.

• *Betaglucanase*

Grapes which have been subject to adverse weather conditions and have suffered a degree of attack from grey mould, or those which have intentionally been subject to noble rot, contain another troublesome substance with a large molecule known as β-glucan. This causes great problems in filtration, as it blocks all membrane filters. It can be eliminated with another enzyme, known a betaglucanase, which has been extracted from another species of fungus, *Trichoderma harzianum*. This treatment is normally done after fermentation.

Another very simple method of overcoming filtration problems when β-glucans are present is to warm the wine to 25°C. At this temperature the β-glucan molecules undergo a change of shape, the viscosity of the wine drops and it becomes filterable.

• *Lysozyme*

Lysozyme is an enzyme found in the protective fluids (tears, saliva and mucus) of most animals, where it fulfils its purpose of killing certain types of bacteria by degrading their cell walls, but is commercially produced from egg whites. Bacteria that are affected are those categorised as gram positive and include *Oenococcus oeni, Pediococcus* and *Lactobacillus*. This last bacterium is where lysozyme has a use in winemaking, as it will prevent, or at least delay, the malo-lactic fermentation. It has no effect on gram negative bacteria such as *Acetobacter*, nor on yeasts.

Because of this effect on some of the bacteria in the wine, the dosage of sulphur dioxide can be reduced.

It is added to black grapes at the crusher to control the bacteria before the onset of the primary fermentation. It can be added to a stuck or difficult fermentation to reduce the risk of increasing volatile acidity due to lactic bacteria. It can also be added after the secondary fermentation, thus allowing a reduced level of sulphur dioxide to be used.

The following two enzymes are not additives, but are present in the grape juice when harvested. They are included here simply for the sake of the convenience of grouping all the enzymes together.

•   *Laccase*

This is an enzyme present in grapes that have been attacked by *Botrytis cinerea*. It is one of the polyphenoloxidase group of enzymes and, as such, it promotes oxidation and causes the wine to turn a deep gold or even brown. This happens even in the presence of adequate levels of sulphur dioxide, and is a particular problem with red wines, as it destroys the red pigments. One way of treating it is to add higher levels of sulphur dioxide than normal, or pasteurisation, which is not ideal for a fine wine.

•   *Tyrosinase*

Also a polyphenoloxidase, but it is present in healthy grape juice and is susceptible to sulphur dioxide at normal levels. It is not a great problem, but is something of which all winemakers should be aware.

Each wine producing country has its own list of permitted treatments and additives. The complete list of those permitted in Europe is to be found in the annexes of Regulation (EC) No 606/2009 and the subsequent amendments, available from HM Stationery Office bookshops, or on-line at:

http://eur-lex.europa.eu/LexUriServ/LexUriServ.do?uri=OJ:L:2009:193:0001:0059:EN:PDF

Chapter 16

FILTRATION

Pretty! In amber to observe the forms
Of hairs, or straws, or dirt, or grubs, or worms!
The things, we know are neither rich nor rare,
But wonder how the Devil they got there.

Alexander Pope 1688-1744

The process of filtration probably causes more controversy than any other single treatment available to the winemaker. There are those who maintain that filtration ruins wine; there is the opposite school that claims that, properly applied, there is no ill effect whatsoever. The fact is that filtration, properly used with care and expertise, should have no noticeable deleterious effect on the wine. Conversely, it is not an essential process for a wine that has been allowed to ferment out to dryness in a natural manner.

Filtration is not necessary for traditionally made wines because they are intrinsically stable once in a hermetically sealed bottle. Yeasts and bacteria will die out due to the combined effects of a shortage of both nutrients and oxygen. Oxidation is prevented provided the cork is sound and the wine will keep until the end of its natural life, the only change being the maturation reactions (see below).

Filtration is a widely varied technique that can be applied for different purposes. Coarse filtration will render a cloudy wine bright. A fine filtration will remove all microorganisms, if necessary. If a wine has been blended in the medium dry or off-dry style, with a small quantity of residual sugar, it will have to be packed in an aseptic manner (see p.216), with a total absence of yeast, to prevent a re-fermentation in bottle. In this instance, filtration is an essential treatment.

The greatest danger with filtration is being over-zealous, using sheets that are too tight in structure or membranes with pore sizes intended for the preparation of intravenous liquids. Properly applied, the sequence of filtrations forms a widely used and very satisfactory system for the bottling of good quality commercial wines. Many of the large retailers insist on a complete absence of all viable microorganisms, which means that so-called sterile filtration regimes have to be used.

Filtration is normally applied in a graded fashion, using coarse filtration first to remove the gross particles, followed by progressively finer stages of filtration, until the wine has been brought into the correct state for bottling. The reason for this progression is that, apart from cross-flow filtration (see p.192), the finer types of filter would be blocked by an excessive quantity of solids.

Principles of filtration

Filtration techniques fall into two main categories, depth filtration and surface, or absolute, filtration.

Depth filtration is so called because the solid particles are removed from the liquid within the structure of the filter medium itself, rather than on the surface. The filter medium has to be many times thicker than the size of the particles being removed because the channels through the filter are mostly bigger than the particles themselves. The filter works because as the particles travel through the tortuous pathways they become trapped somewhere within the depth of the material before they can reach the clean side.

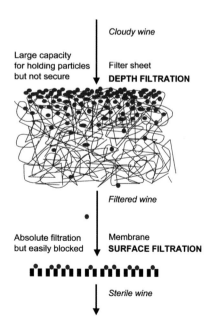

Cloudy wine

Large capacity
for holding particles   Filter sheet
but not secure   **DEPTH FILTRATION**

Filtered wine

Absolute filtration   Membrane
but easily blocked   **SURFACE FILTRATION**

Sterile wine

The advantage of this type of filtration is that it can cope with liquids that are heavily laden with solid particles, such as juice straight from the press, or wine that has just finished its fermentation. The mass of solid particles is held somewhere within the depth of the filter and gradually blocks it, until eventually it has to be discarded.

The commonest forms of depth filtration used in the wine industry are kieselguhr (or earth filtration) and sheets (or pad filtration), both of which are widely used. They are simple to operate and the expendable filter media are of relatively low cost. However, the disadvantage of this type of filtration is that it is not totally safe.

By trying to filter too quickly or by using the filter too long without replacing the filter medium, it is possible to force particles through the filter to the clean side, hence the development of absolute filtration.

Absolute, or surface, filtration gives complete safety, the channels through the filter medium being smaller than the particles being removed. The particles are trapped at the surface of the filter rather than within its depth, hence the name surface filtration. The alternative name of absolute filtration is equally valid because the particles absolutely cannot get through the filter. The disadvantage is that the holes are easily blocked by solid matter which prevents further filtration.

This category of filtration is also widely used in the wine bottling industry in the form of the membrane, or cartridge, filter. Because of its tendency to block very easily it is only used as the final guard filter immediately prior to the filling machine.

Depth filters

• *Kieselguhr filters (earth filters)*

Depth filtration using kieselguhr is the first type of filtration used on wines that are still thick with solid particles from the fermentation. These solids consist of a mixture of dead yeast cells with some live cells, solid matter from the original grape and other solids, which have been rendered insoluble by the effect of the alcohol.

Photomicrograph of kieselguhr as used for the creation of a filter bed

Kieselguhr, or diatomaceous earth, is an earth that is mined in Germany and consists of the skeletal remains of diatoms, tiny sea creatures that inhabited the North Sea many millions of years ago when that part of Germany was under the water. The deposits are mined, ground into a fine powder and treated with acids and alkalis until all that remains is pure silica, which is totally inert. When these particles have been formed into a bed they produce an effective filter by forming a porous barrier whose channels are numerous and tortuous. This substance forms the basis of two types of filter which are used for the first filtration of wine after the fermentation has finished. In both cases the kieselguhr is made into a slurry with water or wine and is dosed into the cloudy wine as it reaches the filter.

Rotary vacuum filter being used for filtering the lees of Marsala wine

The ***rotary vacuum filter*** is the machine of choice for the first stage, as it can cope with liquids which are almost too thick to be called liquid. This filter consists of a large horizontal drum whose cylindrical surface is formed from a fine stainless steel mesh. The lower half of the drum is immersed in a bath containing the cloudy wine. When in operation the drum rotates slowly, the interior of the drum being connected to a powerful vacuum pump. The vacuum draws the wine through the mesh into the interior of the drum from where it is pumped out.

Effective filtration is not occurring at this stage because the mesh of the drum is not sufficiently fine to trap the particles that make the wine cloudy, but it will trap particles of kieselguhr. The bed of kieselguhr has to be built up by re-circulating the wine and adding more kieselguhr. The kieselguhr is gradually deposited as the vacuum sucks the wine through the mesh, and the thickness of the layer increases with each successive rotation of the drum until it reaches several centimetres in depth.

This layer now acts like a sponge, being wetted as it dips into the wine in the trough and being sucked dry as it rotates in the air. True filtration begins at this point, valves are changed over and the filtered wine is pumped into a clean vat. To enable the filter to work continuously, a blade shaves off the top layer of kieselguhr as the drum rotates, thus removing the outer layer of wine deposits and preventing the bed from becoming too thick.

The rotary vacuum filter is used primarily for filtering the lees, which no other filter could handle. It does have the disadvantage that the large surface area of the filter bed is exposed to air with the inevitable risk of oxidation.

The ***earth filter (or kieselguhr filter)*** was developed to overcome this problem. It is totally enclosed and can be flushed with nitrogen before use. Inside, it consists of a series of rotating hollow disks with a mesh

An earth filter that can be wheeled around the winery

surface which operate by exactly the same principle as the rotary vacuum filter.

Kieselguhr can be purchased in a variety of particle sizes, giving filtration of all grades from simple clarification up to yeast removal. Although the filter machine is expensive to purchase, it is versatile and can easily be wheeled around the winery, and kieselguhr is cheap. Such filters are widely used for preparing wine for subsequent stages of treatment and can even remove yeast if the finest grade of kieselguhr is used (sometimes known as *terre rose*, or pink earth, because of its colour).

• *Sheet filters (plate & frame or pad filters)*

Another form of depth filtration used after the gross solid matter has been removed from the wine is variously known as a sheet filter, a pad filter or a plate and frame filter. The last of these is a good descriptive title because the filter consists of a solid, heavy framework with a fixed back plate and a moveable front plate which can be wound in and out by a large screw thread. Between these two end plates is a set of specially designed chambers known as 'plates' which distribute the wine to the filter sheets that are placed between each of the plates in a sandwich formation.

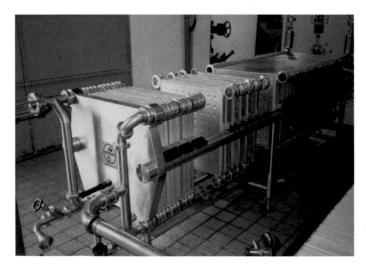

A sheet filter dismantled for cleaning

The important point to note is that the wine does not pass from one end of the filter to the other, through all the sheets, but is distributed by the plates in such a way that each portion of wine passes through one sheet only. The cloudy wine enters the filter on one side (shown blue in the diagram below), is distributed to the sheets, passes through them and collects as bright wine on the other side of the filter (shown red in the diagram). The reason for having multiple sheets is merely to increase the rate at which the wine can be filtered: the more sheets, the greater the flow rate, but the quality of the filtration remains the same.

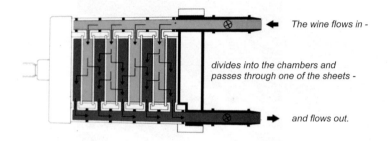

The wine flows in -

divides into the chambers and
passes through one of the sheets -

and flows out.

The mode of action of a plate and frame filter

The sheets themselves are of quite simple construction, consisting primarily of ordinary cellulose fibres, the same as blotting paper. They may contain extra substances to increase their filtration efficiency, such as kieselguhr, but they no longer contain asbestos which was banned in the 1970s as a carcinogen. As with earth filtration, filter sheets depend upon the thickness of the sheet being many times greater than the size of the particles being removed.

It is possible to remove even the microorganisms from wine and render it sterile by using filter sheets of a very fine grade. Such sheets are therefore known as sterilising sheets and are frequently used for removing yeasts and bacteria before bottling the wine. But therein lies a danger because this type of filtration is depth filtration and, due to the structure of the sheet with its mat of fibres and comparatively large interstices, it is possible by mishandling the operation to force yeasts right through the sheet and into the filtered wine. Thus, operators using sheet filtration need thorough training to ensure they understand that the stated maximum flow rate and maximum pressure differential between the inlet and outlet of the filter must not be exceeded.

Despite this drawback, sheet filters are widely used although the initial capital outlay is high.   These machines have to be made of very substantial components to withstand the high forces required to compress the edges of the sheets to prevent leakage.   However, the sheets are inexpensive and are obtainable in all grades from simple clarification (polishing) up to total sterilisation.   Successful filtration depends on the correct selection of sheets for each type of wine.

Photomicrograph of yeast cells trapped on a filter sheet

One of the least attractive aspects of the traditional sheet filter is the dripping edges of the sheets which attract flies and look generally unhygienic.   This has led to the development of totally enclosed versions, where the filter elements are enclosed in a cylindrical stainless steel housing, which greatly improves the cleanliness of their operation. This equipment, incidentally, looks remarkably like a membrane installation, so care should be taken when visiting a winery to avoid making a wrong judgment regarding their filtration technique.

A further development in depth filtration makes use of the fact that many particles in suspension carry an electrostatic charge.   The electrochemistry behind this is very complex and is not fully understood. (The charge on these particles can be expressed in terms of what is known as the *zeta potential*.)   It is possible to incorporate a substance in the structure of the filter medium that also carries an electrostatic charge.   These substances will attract particles with the opposite charge and will thereby remove them from the liquid being filtered.

One of the substances that exhibits this charge effect is asbestos. Filter sheets containing asbestos were widely and successfully used for the final filtration of wine prior to bottling, and great consternation was expressed when asbestos was banned because of its carcinogenic properties. Modern zeta potential sheets contain more benign admixtures, although it has to be said that the old sheets contained brown asbestos rather than the highly dangerous blue variety.

It should be noted that these sheets cannot be used as a substitute for fining despite the use of zeta potential principles; the troublesome colloids are not trapped because they are too small.

Surface filters

• *Membrane filters (Cartridge filters)*

The shortcomings of depth filtration, the clumsiness of the equipment and the uncertainty of filtration efficiency have led to the development of an absolute filter, the membrane. This has quite the opposite characteristics to earlier forms of filter in that it does not rely on a depth of material but removes particles at the surface of the filter. This comprises a thin plastic membrane punctured with minute holes, which are smaller than the particles being removed. (This is a simplistic description of the structure of a membrane, which is actually quite complex, but it will suffice for understanding its function.)

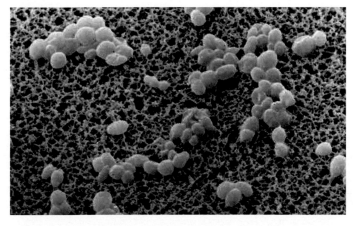

Cells of Saccharomyces cerevisiae sitting on the surface of a membrane filter

The membrane is fragile, so has to be supported on a layer of fibrous material which is then folded into a concertina formation and is sealed into a cylindrical case, looking somewhat like a car oil filter. This is fitted with a rubber O-ring and is plugged into a suitable hole in a solid metal base-plate and the whole covered with a stainless steel housing. The housing is filled with wine and the pressure of the pump forces the wine through the cartridge from the outside to the inside, where it collects and is removed via the base of the unit.

A membrane cartridge cut open to show the construction

This would appear to solve the problems of the removal of particles from wine, but unfortunately there is a drawback. Because these filters have no depth and the filtration takes place at the surface, the holes become blocked very quickly and the flow comes to a total stop. Therefore a membrane filter must be handled with great care and must only be used for the final filtration of wine that is already very clean. Any attempt to filter a dirty wine will result in an expensive change of cartridges.

The costs of membrane filtration are opposite to depth filtration in that the equipment for housing the cartridges is relatively cheap but the cartridges are expensive. However, properly handled, the running costs of membrane filtration are not high because each cartridge will filter thousands, if not millions, of litres of wine without blocking. They are universally used as the final filter prior to the bottling of wine when they should correctly be placed immediately prior to the filling machine to trap any stray yeasts that might have escaped the sterilisation process.

Membranes are manufactured in a range of different pore sizes and an effective way of maximising their life is to use them in sequence, the wine passing from larger pore size to smallest. Alternatively, most membrane filter manufacturers supply what is known as a "guard filter" to protect the final membrane. This filter contains cartridges that look like membranes but are actually made of fibrous material which acts as a depth filter and removes the particles that would clog the membrane. Being depth filters and not absolute filters, these guard filters are given nominal ratings, such '2μ nominal'. This is a useful way of differentiating a guard filter from a true membrane.

A typical filtration rig as used in many bottling halls

The largest size usually encountered in wineries is 1.2μ, which will remove most yeasts but allows bacteria to pass. (μm = micrometre, μ = micron, both of which equal one millionth of a metre) This might be followed by a 0.8μ, which removes all yeasts but does not guarantee the removal of all bacteria. The final membrane is normally 0.45μ, which removes all yeast and all bacteria. It is possible to obtain a 0.2μm pore size, but this is not advisable for wine filtration and should be reserved for the production of sterile water. The danger here is that the pore size is so small that it could remove some of the valuable constituents from the wine itself.

A good principle adopted by some wineries is to relate the pore size to the wine being filtered. The wines which are the most susceptible to microbial damage, such as light German wines containing less alcohol and high residual sugars, need the most stringent filtration. In this case, a 0.45µ pore size might well be used. Conversely, it would be advisable to filter a full-bodied dry red wine through a 0.8µ cartridge to ensure that the body of the wine is not being adversely affected.

• *Cross-flow filters (tangential filters)*

The problem of the short life of a membrane in the presence of solid particles has been overcome by cross-flow filtration, an ingenious modification to the technique of filtration. In all the filtration methods discussed above, the flow of liquid is perpendicular to surface of the filter so all the solid particles collect on the filter, gradually blocking it (or in the case of a membrane, instantly blocking it). By the simple expedient of turning the flow through ninety degrees, so that it is parallel to the surface rather than perpendicular to it, the liquid will flow across the surface of the membrane and will sweep the surface clean rather than blocking it.

PERPENDICULAR FLOW   CROSS FLOW KEEPS
BLINDS MEMBRANE   MEMBRANE SURFACE
SURFACE   CLEAR

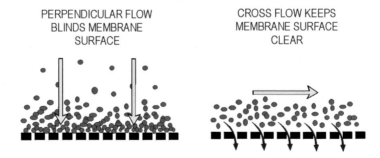

The principle of cross-flow filtration

By arranging for the wine to flow in a complete circuit on the dirty side of the membrane, the wine can circulate repeatedly past the membrane. Under the pressure of the pump, some of the liquid will pass through the pores but the solid particles are held in suspension on the dirty side instead of blocking the membrane. The result is filtration of the wine, with a progressive concentration of solids on the dirty side which can gradually be bled off.

The great advantage of this type of filtration is that the dirtiest of wine, straight from the fermentation vat and heavily loaded with yeast cells and other solid matter, can be filtered to bottling standard in one pass. It is equally useful in the cleaning of must before fermentation.

The disadvantage is that the machine is very expensive, containing many membrane units, either in the form of rectangular cells or formed into fibres which are enclosed in tubes. But this cost must be set against the cost of multiple filtrations using different styles of filter, and the risk to the quality of the wine through multiple handling.

*Sartorius
cross-flow filter
in a modern
winery
in Germany*

Cross-flow filters have gained a bad reputation for stripping wines of body and character, but this is a somewhat unfair criticism. It arose from a misunderstanding of the use of these machines which were originally designed for the production of intravenous liquids for the pharmaceutical industry. To ensure total sterility they used membranes of 0.2μ pore size which is much too fine for wines. These filters give perfectly good results when fitted with the appropriate membrane.

Ultrafiltration

Membranes can be produced with pores much smaller than the smallest in general use, so small that they will actually filter out individual components of wine: the tannins, the sugars and the acids. By the use of this technique, wine can be separated into its essential constituents of sugars, acids, tannins, colours etc., so wine in all its different colours and styles could be produced from one large tank of base wine. This is the ultimate in filtration technique, hence the name of ultrafiltration. It is, however, illegal in Europe (and probably in other parts of the world) and is only used as a research tool, but its very existence poses a threat, if only in theory. What will happen in the future? Shall we need only to produce large volumes of one wine, which can then be manipulated to produce all the different wine styles to order?

Chapter 17

PACKAGING MATERIALS

Nothing in the world was more terrible than an empty bottle! Unless it was an empty glass.

Under the Volcano
Malcolm Lowry, 1947

Containers

There is a wealth of choice in the way in which wine is packed nowadays. Traditions have been slow to break, but no longer are we bound to purchase wine in a glass bottle closed by a piece of natural cork. However, the techniques required for packaging in the various formats demand specialist knowledge because each has its own specific requirements.

The two principles that affect the shelf-life of a package are the size of the unit and the oxygen permeability of the material. Bearing in mind that the changes that occur in a liquid take place at the interface between the liquid and the walls of the container, it becomes clear that the smaller the container, the more rapid the change because the ratio of surface to volume increases with diminishing size of the container. By the same principle it can be seen that a material with a poor oxygen barrier will have a more pronounced effect on shelf-life in a small container.

One point that should be borne in mind before embarking upon the final packaging operation is that European law has stipulated that wine can be bottled only in a certain specified range of sizes (see p.289). The standard size for a bottle of wine is 75 cl, which is helpful when making price comparisons, and this applies to all wines including sparkling and fortified wines.

• *Glass bottles*

Assuming that goat skins are no longer acceptable – or obtainable – it can be assumed that the first container to be considered will be the glass bottle, which is probably the best container for wine that has yet been invented. It is inert and has no possibilities for taint, it is impermeable to gases and is available in almost any shape and size – and even colour. Set against this is the fact that it is fragile and heavy, and is transparent to UV radiation which is a particular danger in supermarkets where bottles are sometimes displayed close to fluorescent lights. (The deterioration due to UV radiation is not oxidation but a breaking down of the components of the wine, which is really a chemical decomposition.)

The techniques of glass manufacture have improved enormously, due largely to the effects of the quality revolution that has taken place throughout manufacturing industry where it is realised that quality is important and that quality sells product. Glass bottles purchased direct from the glassworks are supplied by the pallet, tightly shrinkwrapped in polythene and are virtually sterile, having been palletised whilst still hot. Even the cardboard layer pads between the layers on the pallets have been replaced with plastic to eliminate the possibility of contamination with fibres. It is a pity that many food technologists insist on bottles being rinsed with water before filling because this can actually cause contamination unless the rinsing machine is immaculately serviced.

- ### *Measuring container bottles (MCBs)*

The concept of MCBs was introduced into Europe in 1975 by Council Directive 75/106/EEC and was amended in 2007 by Directive 2007/45/EC. They are bottles whose capacity has been controlled during their manufacture in such a way that the filler of the bottle has only to measure the level of the liquid in the neck to guarantee that, on average, the bottle holds the declared quantity.

The base of an MCB showing the reversed epsilon and the correct filling height: 75 cl at 63 mm from the top of the bottle

The EU Measuring Container legislation puts the onus on the bottle manufacturer to keep the internal volume of the bottles within set limits at a stated level from the top of the bottle. This level is embossed on either the side or the base of the bottle and appears as, for example, 75 cl at 63 mm. The MCB style of bottle is recognised by a reversed epsilon, ᴧ, also embossed on or near the base of the bottle.

When using MCBs, all that is necessary for filling control is to measure the distance from the top of the bottle to the meniscus on a given number of bottles as they travel down the bottling line. The joy of this system is that the test is non-destructive and very simple. It is not necessary (and indeed wrong) to measure the volume of wine in the bottles by emptying the contents into a measuring cylinder because by doing this the useful MCB legislation is being ignored, and the responsibility that belongs to the bottle manufacturer is being shouldered. Nevertheless, many bottlers are still doing it, the reasoning being offered (possibly justifiably) is that they are keeping a check on the bottle manufacturer. It has also been noted that some auditors from accreditation companies are wrongly demanding that this should be done. The International Organisation of Legal Metrology have defined MCBs in R138 as: "Bottles intended to be filled either at constant level or at constant ullage with sufficient accuracy without the need to use an independent measuring instrument." This should be quoted in any instance of dispute.

• ***Plastic bottles***

It is important to distinguish two main types of plastic from which bottles are made: PVC (polyvinyl chloride) and PET (polyethylene terephthalate). PVC bottles are used widely in producer countries such as France, where wine is regarded as an everyday commodity. They are cheap and lightweight but have virtually no barrier to gases; they let in the oxygen from the atmosphere, resulting in a short shelf life. This is perfectly acceptable for wines that are intended for rapid distribution and immediate consumption.

PET bottles are widely used for beers and soft drinks because they have better oxygen barrier properties and therefore offer a reasonable shelf life. However, the UK public has never taken to wine in plastic bottles and their sole use in the wine trade has been in the 18.7 cl size. This quarter-bottle size is popular with airlines for in-flight consumption, where their light weight has proved an advantage. It is also somewhat difficult to transform them into dangerous weapons. They can even be produced with a concave side to prevent them from rolling off the food trays! However, for the dual reasons of their small size and a poor oxygen barrier, the shelf life is only some three to six months and therefore stock control becomes critical.

•   *Aluminium cans*

Market forces have dictated that the only size of aluminium can available for wine packaging is the 25 cl, which makes a convenient two-glass size. The properties of the can are good: strong, light-weight, impermeable to light and to gases, easily filled with the elimination of both air and microorganisms, and a shelf-life of about nine months, and all this with a lower than normal sulphur dioxide level.

The greatest technical problem that had to be solved is the fact that wine and aluminium do not make good companions. The acids in the wine attack aluminium and, in the process, the sulphur dioxide in the wine becomes reduced to hydrogen sulphide, or bad egg smell. This stretched lacquer technology to its limits, but success was achieved and wine in cans has proved to be viable. Their popularity, however, is not great; perhaps because they are more associated with beer or the unit cost is too high, or maybe it is purely traditional resistance to anything that is not glass.

•   *Bag-in-box (BIB)*

The purpose of this style of packaging is to provide a means of purchasing a large quantity of wine that can be drawn off a glass at a time over a long period, with minimum deterioration. The bag-in-box is available in a variety of sizes from 2 litre to 20 litre, the larger sizes being particularly useful for pubs and bars. The technology of this form of packaging is complex and expensive, and it was never intended to be a cheap bulk pack.

The idea of a plastic bag in a box was invented in Australia in 1965 by Thomas Angove, who died on 30th March 2010. But it wasn't until 1967 that Penfolds produced a metallised bag with an integrally welded airtight tap, virtually the same as we see today. It became extremely popular due to the low price – there was no wine duty in those days – and the availability of large quantities of a good quality, easy drinking wine. Another factor in their favour is that Australians possess large refrigerators in which they can keep the boxes. In 2004, even with the imposition of duty, cask wines, as they are called, represented 53% of the total light wine market in Australia. However, this has fallen recently with the decline in sales of lower priced wine and consumers switching to smaller quantities of higher quality wines. In Sweden the proportion of sales of light wine in BIB is similar at around 55%. In the UK the bag-in-box share reached around 12% at its peak during the decade of the 1980s, but it still holds a a significant share of the light wine market.

The principle of drawing off a glass of wine at a time without deterioration has largely been achieved, although the total shelf life of the pack is still limited to about nine months for a three litre box from the time of packing, irrespective of whether or not it has been broached. It is unfortunate that the producers of wines in bag-in-box continue to state: "This wine will keep fresh for up to three months after opening." This is irrelevant, as the critical factor is the age of the filled pack and not the period after opening. It is advisable to purchase bag-in-box wines only when needed, and from an outlet that has a good turnover and can guarantee fresh stock. The shelf life of the larger sizes is longer because of the bigger volume to surface ratio, and they have proved to be a useful way of serving single glasses of wine in pubs and restaurants.

Inside the box there is a flexible bag, which collapses as the wine is drawn off so that air is kept away from the wine. The technology involved is considerable because the material of the bag has to be flexible and yet retain good barrier properties. The tap must not leak and must also prevent ingress of air; and the box must withstand the considerable hydraulic forces of liquid moving from side to side during transport.

The material of the bag has been the greatest technical problem, trying to combine flexibility with a good oxygen barrier. If the bag does not collapse readily, air gets in through the tap when wine is drawn off which results in oxidation of the remaining wine. The obvious choice might seem to be the simple polythene bag, widely used for containing many foodstuffs, inert, non-toxic and easy to weld. Unfortunately, it is as transparent to oxygen as it is to light, and wine stored in a polythene bag will oxidise even more rapidly than in an opened bottle. Most bag material nowadays is a complex structure consisting of two outer layers of high density polyethylene (HDPE) between which is sandwiched an oxygen barrier consisting of a film of polyester which has been coated with a layer of aluminium. Sometimes this barrier layer consists of a very thin sheet of aluminium foil, which gives excellent protection when new but deteriorates when in use due to minute cracks that appear in the foil when it is flexed (known as flex cracking).

Even those bags that look as if they might be made of simple polythene because they are transparent are concealing an equally complex structure. In this case the oxygen barrier is not aluminium but polyvinyl alcohol (PVA), a plastic that prevents the passage of oxygen. PVA does not suffer from flex-cracking, but does not have such a good initial oxygen barrier as aluminium foil and is also very sensitive to water vapour. Therefore the choice is between a better oxygen barrier with potential flex-cracking, and a poorer barrier but more robust.

Filling is also critical and must be more tightly controlled than the filling of glass bottles. The flexible nature of the bag presents the first problem, for the total capacity of a bag intended for three litres is actually nearer to five if filled to the maximum. If not properly supported during filling, a large air bubble could be incorporated in the bag which would be disastrous for its shelf life. The filling machines must be carefully designed and expertly operated to ensure that only the smallest air bubble is trapped in the bag.

The tap is another route by which oxygen can enter. A great deal of research has gone into the design of an ingeniously simple tap. The three basic requirements are that it must allow the wine to flow out quickly when operated, it must not leak when closed and it must not allow oxygen to enter the bag.

The highest standards of wine preparation and microbiological control are necessary. The oxygen barrier has not been perfected, despite the best efforts of the laminate technologists,although it is greatly improved from the early days of bag-in-box. The resultant small amount of oxygen that gets into the bag encourages any yeasts to multiply. Hence the microbiological standards for filling must be a total absence of all yeast and all bacteria, which is not difficult to achieve with modern methods of sterilisation and filtration.

When the wine is packed it is necessary to use a higher level of sulphur dioxide to combat the small ingress of oxygen. This has led to some criticism of the bag-in-box technique, but the sulphur dioxide falls to normal levels by the time the pack is on the retailer's shelves.

A further criticism has been that the quality of the wine in bag-in-box is lower than that in bottles. This was true in the early days when packers of bag-in-box deliberately chose cheap wines to keep the price down. Nowadays there is a large range of good quality wine available in bag-in-box, with a reasonable shelf-life. However, there is a possibility that 'flavour scalping' occurs, with some of the wine flavour components migrating into the plastic material of the bag (see below under synthetic stoppers).

• *Cardboard 'bricks'*

Fruit juices and milk are frequently found packaged in what are generically known as 'cardboard bricks' as exemplified by the Tetra Brik. This is an excellent form of packaging, of low cost, having a good oxygen barrier and a good shelf life. Supermarkets in continental Europe frequently stock the cheaper wines in this format, but it has never been widely accepted in the UK for wine because of its image.

The idea is quite revolutionary and is best imagined as the package being formed around the wine, rather than the wine being put into a package. The structure of the packaging material is important, consisting of a laminate of polythene, cardboard, aluminium foil (the oxygen barrier) and another layer of polythene (to protect the aluminium). The material is delivered on a large reel which is fed into the filling machine, where it is first formed into a tube and the bottom sealed. The wine is piped into the tube from above, the tube is sealed and cut into correctly sized portions. The portions are squeezed into rectangular shape and the ears are bent down to produce the familiar brick shape.

It is the only form of packaging which is truly aseptic, in that the packaging material is sterilised as it passes into the filling machine, the interior of the machine is filled with sterile-filtered nitrogen, and the wine itself is sterile filtered.

Closures

A good closure serves two purposes: it keeps the liquid in the bottle and it keeps oxygen out. These are the first two principles that have to be considered when choosing a closure. Added to these must be the cost, which must be commensurate with the value of the liquid inside the container.

•   *Natural cork*

Cork has characteristics that make it very useful as a bottle closure. It is cheap, readily available, comes from a renewable source, is biodegradable and is a good oxygen barrier. It is elastic, can be compressed, and will quickly regain its original size. It possesses an amazing anti-slip property which holds it in place without undue force from the cork itself. Cork is composed of hollow cells containing air. When cut by the punching tool, the surface of the cork presents a series of sucker-like cups to the bottle neck which holds the cork in place and yet enables it to be removed by applying a gentle force.

Natural cork has been considered the ideal closure for a wine bottle for many centuries. Being a natural product it was expected that its performance would be slightly variable, but this was accepted. Things started to go wrong somewhere around the 1960s. At that time, the world consumption of wine was starting to increase, with most wine supplied in a glass bottle closed with a natural cork. This increased demand for cork put pressure on the producers, who were mostly Portuguese. They started cutting the cork too close to the ground and were generally somewhat casual about their procedures. This situation was not helped by the fact that the large producers and retailers were cutting costs by using corks of a lower grade and a shorter length. The result was a considerable increase in the incidence of bottles with a musty taint, commonly called a 'corked' bottle. It was this that led to the development of alternative closures.

Having been heavily criticised for an apparent increase in this problem, the cork industry rallied some years ago in an extensive project code-named Quercus, by which it admitted to the problem and undertook to take wide-ranging steps towards its long-term cure. It was confirmed that the prime cause of the taint was a reaction between a penicillium mould in the crevices of the cork and the chlorine-containing chemicals used in the sterilising process. This reaction produced trichloranisole (TCA), a substance with a powerful fungal aroma.

In 1996 the European Cork Federation introduced a new Code of Good Manufacturing Practice for Cork which contains various recommendations for improvement. These included careful harvesting of the cork bark, storage off the ground and a replacement of the chlorine process by one based on peroxide, which eliminates the precursor to the TCA. This should result in a considerable fall in the incidence of cork taint, but it is unfortunate that cork has acquired a bad image and the cork industry has been slow in responding. (It should be noted that a tainted cork is not the only source of TCA in wine, see p.254.)

The passage of oxygen through cork is an interesting topic. The process of maturation of wine in a closed bottle used to be regarded as an extension of the maturation that takes place in barrels, where oxygen is an essential element in the process. Then, as knowledge of the maturation process increased, this period in the life of a wine was regarded as anaerobic and merely the slow process of chemical change (see below). Now the situation has been reversed, and with the experience of wine bottled under a hermetic seal, with zero oxygen ingress, it is realised that a small amount of oxygen is needed to prevent the generation of reductive taint (see p.225). So, with an increased awareness of its properties, high quality natural cork will regain its position as the closure of choice for fine wine.

• *Technical corks*

This term covers those closures that are made from cork that has been treated in some way. The simplest and most straightforward of these is the **colmated cork**, which is a piece of natural cork that has been coated with a mixture of cork dust and latex. This treatment fills the lenticels (cracks and crevices) and improves both the appearance and the performance, and is relatively inexpensive.

The cheapest closure based on cork is the **agglomerate cork**, a concoction of cork granules stuck together with a resin-based glue. This is only suitable for wines with a short shelf-life because the resin disintegrates after a few months in contact with wine, leaving the wine contaminated with cork granules. This results in erroneous complaints of 'corked' wine!

The most technical of all technical closures is the **Diam** stopper made by the Oeneo Closures company. The process begins by grinding natural cork into small particles which are separated into the soft elastic fraction (the suberin) and the harder woody particles which are rejected. The suberin fraction is then treated by the Diamant process during which the particles are washed with super-critical carbon dioxide, which removes any residual TCA. (This is the same process used for the production of decaffeinated coffee.) The cork particles are then mixed with microspheres of a special plastic polymer before being compressed and heated to form a uniform, fine textured stopper. The principle of the process is similar to the now defunct Altec stopper, but with the added security of the Diamant treatment.

A very successful modification of the agglomerate principle is the **One-Plus-One** cork (the TwinTop as produced by Amorim), which consists of a disc of natural cork bonded to each end of an agglomerate centre. This clever design has the advantage of a low-cost middle portion protected by natural cork at each end. The wine comes into contact only with natural cork, and the agglomerate centre is protected from attack by the wine.

• *Synthetic closures*

A vast amount of research has gone into the production of an artificial replacement for cork so as to avoid the possibility of TCA taint. The synthetic stopper is made from a type of expanded plastic and has proved to be very controversial. The early versions were difficult to apply, causing damage to corking machines, and were equally difficult to remove from the bottle. Some would claim that traces of chemicals enter the wine from the plastic and that the plastic itself absorbs flavour components from the wine. So what are the facts?

Firstly, to eliminate any fears of nasty chemicals leaching into the wine, the stoppers are made from the same plastic as used in the manufacture of components for surgical purposes. Many foodstuffs nowadays are packaged in some form of plastic container, so it is irrational to level this complaint against the use of a plastic wine stopper. After all, some wine is actually packaged in plastic bottles.

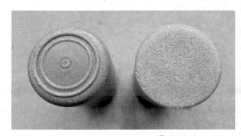

Moulded synthetic stopper   *Extruded synthetic stopper*

The manufacture of synthetic closures falls into two categories, the moulded and the extruded. Moulded closures can be recognised by slightly domed ends, with mould marks around the perimeter. These were the first type to be produced, typified by SupremeCorq. The earlier versions tended to be rather hard, difficult to remove and almost impossible to replace, although more recent productions show improved flexibility.

Closures of the extruded variety are easy to identify, looking somewhat like a piece of plastic pipe filled with foam, which is precisely what they are. They are more elastic than the moulded variety and are therefore easier to remove (and replace, if you must). Two well-known producers are Neocork and Nomacorc.

One of the disadvantages of stoppers made from any form of plastic is that they do not have a good oxygen barrier, the same problem as is faced by suppliers of bag-in-box wines. This is not a great concern for wines for everyday drinking and with a relatively short shelf-life, but producers of fine wine intended for laying down are rightly very cautious about using synthetic stoppers.

Another problem is that of 'flavour scalping', the absorption of flavour compounds from the wine. These substances are hydrophobic, meaning they do not like being in an aqueous environment but prefer to be in a more oily situation. This is precisely the nature of plastics materials – water will not wet them, but runs straight off. So the flavour compounds leap out of the wine and into the plastic material. This can be confirmed by smelling the strong wine aroma on a synthetic closure after removal from a bottle. The effect on the wine is probably small, but nevertheless, it exists.

Synthetic stoppers can be produced in a huge range of colours and styles

There are those who would say that it is a pointless product – it is trying to imitate natural cork, which itself has imperfections and one cannot help wondering why manufacturers should bother with the production of such a closure. The answer is simple: everyone likes the ceremonial of the extraction of the cork, with its accompanying "plop!", and nobody likes the taste of TCA. And it is available in a range of bright colours, which delights marketing departments!

• *Aluminium screwcaps*

Screwcaps have become fashionable, having endured years of consumer resistance because of an association with cheap wine. They are now regarded as 'cool'. In Australia they are known by the acronym ROTE (Rolled On Tamper Evident), and in the UK by ROPP (Rolled On Pilfer Proof), and around the world the brand name of Stelvin is well known.

On the positive side, they do not cause taint, should not suffer from quality variation and do not need a special tool for opening. Trials conducted some twenty years ago showed that the metal cap provides the best means of closure for glass bottles. More recently, tests have been ongoing with Clare Valley Riesling wines *(et al.),* showing excellent development of tertiary aromas. Many of the large retailers in the UK are asking for this type of closure on virtually all of their wines, and many new world wine producers are switching over unilaterally.

They undoubtedly have the best oxygen barrier of all the closures, and should give a perfect seal every time, provided the capping machine is correctly adjusted. The thread on the caps is formed by compressing the metal against the bottle so that it conforms to the shape of the threaded neck. If the capping machine is badly adjusted, it produces caps that refuse to unscrew. Also, because the aluminium from which these caps are made has to be rather soft, they are liable to physical damage during storage or transport, leading to an imperfect seal that allows oxygen to enter the bottle. Recent developments have lead to a revised profile to the top of the cap which has all but eliminated this risk.

Some winemakers are worried by the totally hermetic seal. The old school of winemaking held that small amounts of oxygen were necessary for the proper development of red wine in bottle, and this is becoming accepted more and more as experience with screwcaps increases. There can be no doubt that bottling under this type of closure requires a reduction in the level of sulphur dioxide to compensate for the smaller quantity of oxygen entering the bottle. Failure to do this can result in the formation of hydrogen sulphide, otherwise known as 'bottle stink'. There is evidence that the seal of the original screwcap is too good, even when wine is bottled with the greatest attention to detail.

It is ironic that the closure that was developed specifically to form the perfect seal against oxygen is now available with a choice of three different wads (seals) in the cap: the original wad with an impervious layer of tin, and two grades of Saranex (a polyethylene plastic) which deliberately leaks oxygen. So, the old school of winemakers was right after all!

Caution should be exercised in any decision regarding the use of screwcaps. They are undoubtedly excellent for wines for everyday drinking, whether white, red or pink, but the cork probably still reigns supreme for wines for laying down and wines matured in oak, but it has to be of superfine quality and full length and is therefore expensive.

• *Glass stoppers*

This is a relatively new closure, frequently known by its brand name of Vino-lok. Unlike the old ground-glass stopper, the seal is formed by a PVC plastic ring that sits at the top of the stopper. Specially produced bottles have to be used because the tolerance of the diameter of the inside of the neck has to be smaller than in ordinary bottles, to ensure a tight fit. The stoppers are produced by a special moulding method that includes 'fire polishing' to give a tough finish that minimises the risk of chipping during the wine bottling procedure. The stopper is held in place by an aluminium cap. The advantages are that the oxygen ingress is very low and there is no risk of added TCA, the disadvantage being cost, which is equivalent to a high quality natural cork.

*Vino-lok stopper
as used by Royal Somló
for their Juhfark wine*

• *Other closures*

Several other proprietory closures have been invented, notable among which is the Zork, a plastic device which is 'Easier to get off!' as it has an integral tear strip and does not require a corkscrew.

Capsules

The traditional capsule was developed from the practice of protecting the exposed end of the cork to prevent it from drying out. The earliest way of doing this was to dip the neck of the bottle in sealing wax or beeswax, a practice that is still occasionally seen, if only as a marketing tool.

• *Lead foil*

The first capsules were made of lead foil which served the dual purpose of keeping the top of the cork from drying out and, due to the toxic nature of lead salts, of killing any cork weevils that might try and enter. With the increase in scientific knowledge, it was realised that this practice might also have a detrimental effect on the drinker if the wine became heavily contaminated with lead.

• *Pure tin*

Pure tin is the material that gives the best appearance but it is also the most expensive. It is malleable and easily moulded to the neck of the bottle by a machine known as a 'spinner'. This has a series of small wheels that spin round the top of the bottle, moulding the capsule to the contours of the neck. The appearance is high class, it cuts easily for opening and it is non-toxic, but it is very expensive.

• *Tin-lead*

In order to overcome the danger of wine being contaminated with lead, the tin-lead capsule was developed. This was a sandwich of lead between two layers of tin, an attempt to keep the cost of the capsule down. This material stayed in use for many years, until it was realised that the outer tin layers were usually imperfect, allowing wine access to the lead in the middle of the sandwich, resulting again in the production of lead salts. After a major research programme by the UK Ministry of Agriculture, Fisheries and Food (MAFF), it was discovered that the first glass poured from a bottle whose neck was encrusted with lead salts contained sufficient lead for it to be rendered a toxic drink. As a result of this research, all capsules made from any material containing lead were banned in all European countries from 1993.

•   *Aluminium*

The only other metal capsule that is acceptable is that which is made from aluminium. This metal is not as malleable as tin and requires careful application if creases are to be avoided, but it is cheaper.

•   *PVC*

The most widely used material on everyday wines is PVC, which is available in a wide range of colours and finishes and presents a very acceptable appearance, indistinguishable from expensive tin until the time comes for opening. These capsules are made from an ingeniously produced PVC that has been stretched in one direction during manufacture. After applying the capsule to the bottle, it passes through a heat tunnel where the plastic is softened and thereby shrinks back to its natural size, giving an excellent fit to the bottle neck.

•   *Polylaminated*

Also available is a polylaminated capsule, made from a sandwich of aluminium and LDPE (low density polyethylene – polythene). These are 'spun' on to the bottle in the same manner as tin capsules and produce a good finish.

Chapter 18

STORAGE & BOTTLING

Neither do you put new wine into old wine-skins; if you do, the skins burst, and then the wine runs out and the skins are spoilt.

Matthew 9 [17]

The long process of transformation from grape to wine has been completed. The final blend has been made; the analysis has been done; the wine has been stabilised. All that remains is to package it into suitable containers for transportation to the place of consumption. But the technology behind this chapter in the life of the wine is considerable. A fine wine can easily be ruined by careless handling, by allowing oxygen to dissolve, by allowing bacterial attack, by contamination with iron and copper, or with cleaning chemicals. Fortunately for the drinker, pathogenic microorganisms cannot survive in wine due to the presence of alcohol and the natural acids.

Storage without change

When a wine has been made and the treatments completed it has to be stored somewhere safe where the changes will be as small as possible. There are four conditions that need to be satisfied at this stage, especially for high-volume branded wines, whose character should not change from the beginning to the end of each year and from one vintage to the next.

- The material of the vessel must be impervious: it will not allow oxygen from the atmosphere to enter the wine.
- The material of the walls must be inert: there must be no flavour exchange between the wine and the walls of the vessel.
- The vessel must be as large as possible.
- The vessel must be brimful, or blanketed by an inert gas.

In short, the wine must be put into the largest available vat that is made of impervious, inert material, brimful or blanketed with inert gas.

The choice of construction is not important, provided it is in good condition. A stainless steel vat is probably the first choice in a modern winery. Concrete vats would be equally good provided they are lined

*A fine range of stainless steel
storage vats at
Reichsgraf von Kesselstatt
in the Saar region of Germany*

either with epoxy resin or with tiles in good condition. Mild steel vats are just as satisfactory provided they are epoxy-lined or enamelled.

Vats come in all shapes and sizes from 25 hl (2500 litres) up to 500 hl (50,000 litres) and possibly even bigger. A modern, well-equipped winery should plan to have a selection of vats of different sizes so that every batch of wine can be held under optimum conditions. Ideally, wine would be stored in the largest available vat provided there were sufficient wine to fill it to the brim. An interesting phenomenon known as the 'surface effect', which applies to any container from vat to bottle, is that the smaller the container, the greater the change in the contents over a given period. This is because chemical changes occur at the interface between the liquid and the surface of the container, and the smaller the container the greater the ratio of surface to volume.

If it is not possible to keep the vats brim-full, the headspace must be oxygen-free; the air must be replaced with an inert gas, preferably nitrogen, which does not dissolve in the wine. Carbon dioxide, being heavier than air, is very good for sweeping the air out of empty vats but it has the disadvantage of being soluble in wine. If left above the surface of wine it gradually dissolves, especially if the wine is cold, letting air into the headspace of the vat. Conversely, if wine is kept under pure nitrogen for a long time, the residual carbon dioxide in the wine which gives it a 'lively' taste is lost and the wine becomes somewhat flat. The ideal gas for blanketing purposes is a mixture of nitrogen and carbon dioxide, the proportions depending on the desired carbon dioxide content of the wine (see p.34).

Because of the constant loss of sulphur dioxide as it does its work, analysis of the free sulphur dioxide level must take place after every movement and every treatment, and weekly during storage, with appropriate additions as necessary.

The final sweetening

Wines for sale as commercial blends frequently contain added sugars so that they can be sold as 'Medium Sweet White', for example. Even red blends often contain small amounts of sugars to soften the mouth-feel and to render the wine palatable to those who normally claim not to like red wines.

These sugars are not truly 'residual' sugars, as they do not originate in the grape juice from which the wine was made. They are sugars that have been added to the finished wine in the form of unfermented grape juice, probably as rectified concentrated grape must (RCGM). If it is a German wine at QbA level, then the sweetening will take the form of *Süssreserve*, unfermented grape juice that has been preserved specifically for this purpose, probably by refrigeration.

This sweetening operation is left until the last moment before bottling because, once made, the wine becomes very susceptible to attack by microorganisms, especially yeasts which will start a second fermentation.

It should be noted that European regulations forbid the addition of sucrose (ordinary sugar from sugar beet or sugar cane) for this purpose. The only occasions on which sucrose can be used is for enriching grape must prior to fermentation (see p.65) or for the final sweetening of champagne in the form of the *liqueur d'expedition.*

Shipping in bulk

In the past it was the generally accepted principle that the best way of assuring the quality of the wine was to have it bottled in the country of origin, if not in the actual winery. As attitudes have changed and we have become more aware of the 'carbon footprint' of products around the world, it makes good sense to bottle the wine in the country of sale. Bulk transport has become a highly specialised industry with good standards of quality control throughout the process. Retailers have realised that they can have more influence over the bottling process when it takes place within their own country. Stainless steel ISO tanks can be used for continental journeys, but single-use 'Flexitanks' are becoming popular for deep-sea transport. A few basic steps in quality assurance need to be taken to ensure that the required standards are maintained.

1. Visit the supplier and carry out a quality audit.
2. Agree a specification and a regime of procedures.
3. Engage a reliable transport contractor.
4. Insist on pre-shipment samples of the wine being sent.
5. Insist on loading samples being available if necessary.
6. Ensure there is a strict regime at the reception point.

Bottling processes

Fortunately for the wine trade, wine is a very easy substance to handle. Pathogenic bacteria cannot survive the low pH combined with the moderate level of ethanol found in most wine, irrespective of how it was made. The only requirement for the bottler is the maintenance of quality and the avoidance of foreign matter. HACCP studies (see p.270) have shown that one of the principal critical control points is the prevention of pieces of broken glass and flying insects from entering the open bottles.

Traditional bottling

Wine was bottled without any form of treatment long before aseptic techniques were discovered. Given time and a good winemaker, wine is naturally self-clarifying and can be bottled without fining or filtration, provided it has been completely fermented out and the cork is sound. The wine will come to no harm because the yeast will die out, the alcohol will prevent the growth of any harmful organism, and the acetobacter will have no effect due to the absence of oxygen.

There could be the occasional contamination by bacteria from other sources, and a porous cork might allow volatile acidity to develop, but usually the wine will be able to mature in the bottle to complete satisfaction. It is becoming fashionable to offer such wines for sale, when they are given fancy titles such as *'Cuvée non-filtre'*.

In this method of minimalist technology, the wine is fined if necessary, racked to render it bright and then bottled with no further treatment. It is quite likely that some deposit will form in the bottle after a few weeks or months, but this will probably not matter to the type of consumer that is likely to drink this style of wine, and is used to decanting vintage port and fine claret.

'Sterile' bottling

Modern bottling is often referred to as 'sterile' bottling, which is not correct because sterile means the total absence of microorganisms. This is not true of wine bottling because no attempt is made to eliminate all microorganisms. Indeed, it is not necessary to go to such lengths, even with the most delicate of wines.

Aseptic bottling is an alternative term that is often used. This would seem to be a satisfactory description, as the word 'aseptic' means the prevention of putrefaction, which is precisely the object of the technique and involves the removal of the yeasts and bacteria which cause re-fermentation or volatile acidity. However, some bottling technologists do not accept the use of this term, preferring to call it 'hygienic bottling'. It would be hoped that all forms of bottling are hygienic, so this description is not satisfactory.

The fact is that there is no unique description of this style of bottling that satisfies both scientists and technologists. The nearest one can get is simply 'modern bottling'.

Principles of modern bottling

If a liquid has to be packed in a sterile condition, as with intravenous injections, there must be a total absence of microorganisms. Such strict conditions are not necessary for the filling of wine containers. The critical factor is the microbiological loading, meaning the population density of the microorganisms. Below a certain figure, the wine is safe because the organisms will die out; above this figure they will multiply and the wine will be spoiled.

The only truly aseptic wine packaging is the 'cardboard brick' type of package, where the packaging material is sterilised by UV radiation and the filling takes place in a sterile enclosure.

With ordinary bottle filling machinery, no attempt is made to sterilise either the enclosure or the air entering the filler bowl; these are merely kept hygienically clean. However, the interior of the filler itself, the tanks holding the filtered wine, the pipework and the filters are all thoroughly sterilised prior to each operation. The purpose of this is to eliminate any breeding colonies of microorganisms and to keep the numbers of viable organisms as low as possible.

Sterilisation is best achieved using a copious flush of hot water at 90°C for at least 20 minutes. This has the dual effect of cleaning out any residues and killing microorganisms at the same time. Steam is sometimes used for this operation but is less satisfactory, as it lacks the cleansing power of hot water and can result in the baking-on of residues if the system has not been thoroughly flushed with water prior to steaming.

Chemical sterilisation is another option. The currently favoured sterilant is peracetic acid, whose breakdown products are the non-toxic hydrogen peroxide and acetic acid. This chemical can even be left inside equipment, acting as a residual sterilant.

Chlorine-based sterilants, such as hypochlorite, are no longer regarded favourably because they are sources of chlorine which could lead to the formation of the dreaded TCA.

Dimethyldicarbonate (DMDC)

Although used in Germany for several years, DMDC has only recently been approved for use throughout the EU. Its action is somewhat like sorbic acid in that it is an aid to the bottling of wines with residual sugar. However, its mechanism is quite different, as it acts as a sterilant by inactivating the enzymes in the microorganisms, effectively killing them.

DMDC is a colourless liquid that is added to the wine just prior to bottling. Its useful attribute is that it is hydrolysed (broken down) by water to methanol and carbon dioxide after about four hours at 20°C, both of which are natural components of wine, so it does not have to be declared as an ingredient. It is purely a processing aid.

Modern bottling techniques

With the advent of modern technology it has become possible to stop the fermentation before reaching its natural conclusion. Such wines are unstable and will go into a second fermentation the instant a stray yeast enters the wine. Further, to satisfy the demand for medium dry wines, unfermented grape juice is frequently added during the final blending process, making the wine even more attractive to yeasts. The only way such unstable wines can be bottled is by the use of 'aseptic' bottling techniques. These techniques have become the norm in many modern wineries, where all wines are bottled aseptically even when it is not

Modern totally enclosed filling machine

necessary. It could be said that technicians have had too strong an influence on the bottling process, demanding levels of sterility that are quite unnecessary and possibly even detrimental to the quality of the wine. As always, minimum treatment should be the rule.

Yeasts and bacteria can be eliminated either by heating the wine, which kills them, or by filtration, which physically removes them. Heat can damage wine, so it needs to be applied carefully, with knowledge and with strict quality control procedures, and of course, heating involves energy and energy is expensive.

There are three principal ways in which heat can be used, with different combinations of time and temperature: a high temperature for a very short time as in flash pasteurisation, a medium temperature for a medium length of time as in tunnel pasteurisation, or a low temperature for a long time as in thermotic bottling.

All of these methods are controversial because there is one school of opinion that insists that any form of heat applied to wine is damaging. Despite any emotive arguments about heating wine, from a scientific point of view it is accepted that heat will affect the way a wine matures because reactions proceed more quickly at elevated temperatures. Maturation is all about chemical reactions between the constituents of the wine. However, this is all a matter of opinion, and very controversial. There is a well known and highly respected producer in Burgundy who flash pasteurises all his red wines simply because he considers that, for his wines, this gives the best result and is preferable to filtration. (The technical advisor to the founder of the company was Louis Pasteur, so it is hardly surprising that the company still uses pasteurisation!)

• *Bottle rinsing*

It is common practice with all modern bottling establishments that all bottles are purchased new direct from the bottle manufacturers. The cleaning and sterilising of used bottles is not practical, neither from a production angle nor from a financial consideration. The difficulty in ensuring that the bottles are not contaminated with paraffin or petrol or a host of other domestically used liquids make the operation untenable.

Bottles purchased shrink-wrapped direct from the glassworks are clean and virtually sterile, as they are packed whilst still hot. In theory they should be ready for filling. It would seem to be a pity that most supermarkets insist on bottle rinsing, as this operation can easily result in a less clean bottle unless the machinery is maintained in immaculate condition.

*Bottle rinser
at Sichel's
bottling plant
at Cenac*

If bottle rinsing machines have to be used, the rinse water should be fresh and not re-circulated, and should either be filtered through a 0.2μ membrane or treated with sulphur dioxide or ozone to destroy any microorganisms that might be present. The most meticulous operators use both chemical treatment and filtration. A well-designed rinsing machine has a sector after the jetting where the bottles are held at an angle to facilitate thorough drainage, although some establishments rinse with filtered wine to avoid the tiny trace of rinse water that is inevitably left in the bottle.

• *Thermotic, or Hot Bottling*

This method of bottling makes use of the lowest level of heat with the longest cooling time. The wine is heated before going to the filling machine by passing it through a heat exchanger which heats it just sufficiently to bring the temperature to 54°C. This is far below sterilising temperature, the standard conditions for which are 82°C for at least 20 minutes, conditions which are used, for example, for the sterilisation of hospital instruments. It is bottled at 54°C, corked, packed into cases and put away in the warehouse, where it cools slowly to ambient temperature. The yeast and bacteria are killed by virtue of the long period at elevated temperature, rather than the temperature itself.

In the early days of this technology, it was said that this method damaged the wine because of the effect of the heat. This was probably true in some cases, and, indeed, one almost expected the cheap one-and-a-half litre bottle with the screw cap to taste as if it had been used for boiling cabbages. But times have changed, and with good temperature control (and it has to be very good with a variation as small as ±0.2°C) this method produces results possibly even better for some wines than by cold filling methods. This is especially true of young, simple wines, where the small degree of heat advances the maturity slightly, giving the wine a softer and rounder character.

One great advantage of this method is that the contents of the bottle are sterilised after filling and after the bottle has been closed with the cork or the cap. Any yeasts or bacteria that are present in the wine or on the bottles or corks will be killed. The benefit that this offers is that it is not necessary for the operators on the bottling line to adopt sterile techniques when the wine is being bottled. On the other hand, it does not give *carte blanche* for operating to low standards. It is still necessary to ensure that the microbiological loading is low because none of the methods involving heat can cope with large numbers of microorganisms.

The one practical problem that has to be overcome is the control of filling level. If the wine is adjusted to the correct filling level when hot, it will be under-filled when at room temperature. Allowance has to be made, therefore, for the contraction of wine when filling takes place at an elevated temperature.

• *Tunnel pasteurisation*

Louis Pasteur, the great French microbiologist, discovered that heat treatment slowed the deterioration of both milk and wine. The process of pasteurisation has been adapted in two forms for the bottling of wine. The earlier of the two processes, known as tunnel pasteurisation, involves heating the wine in the bottle, after bottling and capping. Like thermotic bottling above, this process has the benefit of requiring only good hygiene during its operation, not sterile techniques, all the organisms being killed in the closed bottle.

The difference from thermotic bottling is that for tunnel pasteurisation the wine is bottled cold and the bottles of cold wine are passed slowly through a large heat tunnel where the contents are raised to 82°C for about fifteen minutes by passing through sprays of hot water. Compared with thermotic bottling, a higher temperature is used, but for a shorter time. At the end of the tunnel the bottles pass through cold water sprays which bring the temperature of the wine quickly down to room temperature before any damage can be done, the whole cycle taking about forty-five minutes.

Bottles inside a tunnel pasteuriser being sprayed with hot water

The equipment is expensive to install because it involves the installation of a large heat tunnel fitted with conveyors, water sprays, pumps and heaters. It also uses expensive energy, although modern tunnels make use of heat exchangers between the various stages of heating and cooling.

Accurate temperature control is paramount if the quality of the wine is to be protected. Criticisms similar to those of thermotic bottling could be levelled at tunnel pasteurisation if it is carelessly operated. Yet this technique is widely used, even for sparkling wines where one might expect the internal pressure to shatter the glass. A producer of Asti (the new name for Asti Spumante) was asked why he used tunnel pasteurisation as well as sterile filtration, and the answer given was: 'So that I can sleep at night!'.

• *Flash pasteurisation*

A flash pasteuriser is a heat exchanger that is capable of raising the temperature of a liquid rapidly to a high temperature, and then bringing it down again as rapidly, literally 'in a flash'. Pasteurised milk is treated by flash pasteurisation, which has been standard practice for years as a protection against the many harmful organisms that milk can carry. It was a simple step to adapt this method for the packing of wine. Wine is intrinsically a much lower risk substance than milk, so it is not necessary to use such stringent conditions.

A pasteurising unit that can be used for hot bottling or flash pasteurisation

For milk the temperature is usually between 80 and 90°C, but it is held there for only a few seconds before it is cooled down again to prevent any hint of cooking. Conditions used for wine are generally around 75° for up to 30 seconds.

The equipment for this type of bottling is simple, consisting of a metal plate heat exchanger placed between the wine vat and the bottling machine. The wine is heated and cooled before reaching the filler, which effectively knocks out all the yeasts and bacteria. However, this simplicity conceals a potential problem, the wine being sterilised before being bottled. It is easily re-infected by the bottling machinery and the bottling materials, in fact by anything with which it comes into contact.

This necessitates the operation of the process by people who have been trained in sterile techniques, people who understand that yeasts are in the air around us, on all surfaces, on our hands. The filling machinery and pipework must be sterilised before use. Bottles should be freshly purchased from the glassworks and must be received packed in good pallets, tightly shrink-wrapped. They should not be stored for longer than a month in case they become contaminated. Corks must be purchased in sealed plastic bags, sterilised either by sulphur dioxide or gamma radiation.

All of these conditions are relatively easy to manage, provided everyone involved is given adequate training. Flash pasteurisation and cold sterile filtration have become widely used techniques in the wine trade.

• *Cold sterile filtration*

If heating is to be avoided, the only alternative process for bottling is to use filtration techniques to remove the microorganisms, a technique known as cold sterile filtration. It is known as 'cold' because the wine has not been heated, not because it has been chilled.

Filtration for aseptic bottling has become probably the most widely used technique of all because it is simple to operate, cheap and reliable. It does have its antagonists who claim that filtration removes the heart and soul of the wine and should never be used. In reality, it depends on the way in which it is used: if used intelligently and sensibly, the effect on the wine is imperceptible. Comparative tastings have sometimes taken place immediately after filtration, when a difference has been detected. The only fair test for the effect of filtration is to allow the wine to rest a few weeks in bottle before tasting.

When a wine has reached the bottling stage, it should have received its initial filtration through kieselguhr and should already be clear and bright. It might still, however, contain low levels of yeast and bacteria which will be removed by the next filtration, which should be sheets of a tight formulation. This is the stage where the greatest damage can be done, by the over-zealous use of filter sheets that are of too dense a structure and can actually strip the wine of body and flavour. Sheets are available in a wide range of grades, which enables the choice of the correct structure to be made relative to the wine to be filtered.

Before the days of the ubiquitous membrane filter, sheet filtration was regarded as the final treatment, being perfectly capable of removing all the yeasts and the bacteria, provided it is correctly operated. It is this caveat that has seen the tremendous rise in the use of membranes, for they offer the security and peace of mind that all bottling operators seek, so that they can sleep at night in the confidence that the highly susceptible wine that they have just bottled will remain stable until opened for consumption.

The final membrane, which should always be situated as close to the filling machine as possible, takes out any stray yeasts that might have got through the sheet filter or that might have survived the sterilisation process in the machinery. Membranes are extremely tough, are easily sterilised, and will hold back anything larger than the size of the pores, but care is still needed, as dirt will easily block them and replacements are expensive. A large, wealthy winery could, of course, use a tangential filter, which would eliminate all the other forms of filtration (see p.192).

The crucial parameter at this stage is the choice of pore size. The zealous technologist must be curbed by the winemaker and encouraged to use only the pore size that is necessary for the removal of troublesome microorganisms. For a big and fat red wine it might be possible to dispense with the membrane altogether. A lighter red wine might need a 1.2μ to remove only the yeasts, whereas a rosé or a white could tolerate a 0.65μ or even a 0.45μ to no detriment. Using a 0.45μ membrane for all wines makes life simple, but does pose a risk to the quality of heavier red wines.

The critical control points for success are the following:

1. Ensuring that all bottles and closures are purchased in a sterile condition.
2. Training the operators in aseptic techniques: an awareness that we are surrounded by microorganisms that must not be allowed to contaminate the bottled wine.
3. Sterilising the filling machinery and testing for the absence of yeasts and bacteria.

4. Filtering the wine carefully through kieselguhr, then sheets, and finally a membrane.

5. Filling the bottles and corking them quickly.

After filling, samples should always be tested for the absence of yeasts and bacteria before releasing for sale. Even the most well organised operation has occasional failures, but a product re-call is very expensive. The cost of re-bottling is much less if the product is still in the warehouse.

Maturation in bottle

After bottling, the final stage of maturation takes place. Bottle ageing is a distinctly separate phase in the maturation process in that the conditions are anaerobic: oxygen does not have a part to play. Or does it? The old idea that the natural cork in the bottle transmits the perfect amount of oxygen to aid maturation may be correct, and perhaps cork *does* play a positive role in the maturation process. This is one of the many areas of wine technology where there is a need for more research. The evidence from screwcap bottlings is beginning to show that a little oxygen *is* needed to prevent the development of reductive odours.

Tokaji Aszu maturing in the cellars at Chateau Pajzos

Nevertheless, it is known that the reactions taking place are mainly chemical reactions between constituents of the wine, largely between alcohols, acids and water. Alcohols and acids form esters, which in turn are broken down by water to alcohols and acids, which then inter-react producing yet more esters, and so on.

Over 500 different compounds have been identified in the aroma of a mature red wine.   These include alcohols, volatile acids, esters, aldehydes, ketones and all the products of interaction, decomposition and hydrolysis (their reaction with water).

The 'surface effect' as described above has a big influence on the rate of maturation in bottle, which explains why magnums mature slowly, while half-bottles reach their peak fairly quickly. The ultimately better wine in a magnum is undoubtedly due to the slower rate of chemical reaction of the constituents, which gives more opportunity for inter-reactions to take place. This results in a greater variety of chemical compounds and hence a greater complexity.

During this period of anaerobic maturation the wine needs to rest under constant conditions. Rapid changes in temperature upset the slow process of chemical inter-reactions; light, and especially UV radiation, speeds up the decomposition of the sensitive components; vibration prevents the settling of the fine deposits that are produced. All of these phenomena confirm that the tradition of keeping bottled wine in a cool, dark, dry cellar is important if the best results are to be obtained.

However, there is no need to despair if such a cellar is not available for a personal collection of wine. All that is necessary for reasonable storage is darkness and a temperature that does not fluctuate violently: the actual temperature is less important, but the cooler the better. A cupboard under the stairs might suffice, or alternatively, a solidly built cupboard at the back of the garage. The worst conditions would be in a loft that gets hot during the day, and dips down towards freezing each night.

Chapter 19

QUALITY CONTROL & ANALYSIS

Wine has two defects: if you add water to it, you ruin it; if you do not add water, it ruins you.

Spanish Proverb

Analysis is an essential element in the overall control of the quality of wine, but quality control is much broader than mere analysis. Quality control should be the responsibility of everybody who has any input to the creation of the final product. In the case of wine, the responsibility lies with the person who tends the vines, the people who pick the grapes, the tractor driver, the press operator, the winemaker, the filling machine operator, the packer, the warehousemen and anyone else through whose hands the wine passes. They all play their part in safeguarding the quality of the wine.

Automatic analysers are gradually replacing the traditional laboratory

Laboratory analysis is the means by which the necessary controls and processes can be selected and applied. Analytical methods are available for all the important components of wine and range from the simplest methods that can be operated in a corner of the winery to those requiring a research-style laboratory with gas chromatograph, atomic absorption spectrometer, etc. A good winery must have properly defined quality control procedures, one of which should be a regime of analysis, taking due note of the requirements of consumer countries. The modern laboratory can be fully automated, producing results which are fed directly into a computer, even controlling the process by a feed-back loop.

Quality plan

A scheme of analysis should be developed that produces useful and meaningful results, not simply those analyses that are easy to perform. A well-designed analytical scheme plays a large part in the production of healthy wine with a good shelf-life. There is little point in filling books or computer spreadsheets with numerous data that serve no purpose.

A regime of analysis should be devised from the time that grapes are being studied for approaching maturity, so that all the important parameters are tested at the appropriate moment. This plan should extend throughout the processing of the grapes, through the production of must and the transformation into wine, to the storage in bulk and the final packaging into the ultimate container. And a good quality plan would not finish here: it would extend into post bottling studies of shelf-life and the analysis of any sub-standard returned bottles.

Records and traceability

The recording of results is of the utmost importance to a winemaker, for it is only by the recording of all the facts that it becomes possible to determine the reasons for success or failure. By studying these facts it should be possible to achieve continuous improvement. In a well-designed scheme of analysis and recording, it should be possible to trace the history of a wine from the lot number on the bottle, back through the blending to the individual components, and possibly even to the grapes from which each component was made. This not only gives useful information for the purposes of continual improvement, but also gives good protection in any case of complaint or investigation. It is nothing more than common sense, and yet some wineries entrenched in their old ways cannot see the logic in this approach!

Records should be an integral part of the quality plan and should follow the passage of the grape through to its sale in bottle. These should be designed in such a way that immediate retrieval is possible. They can range from the simplest set of tables kept in books or files, to the most sophisticated computer based records where complete traceability is possible at all levels.

Traceability is required in Europe by Regulation (EC) 178/2002 which states that "The traceability of food . . . shall be established at all stages of production, processing and distribution.". The effect of this is that all wineries will have to install systems that enable traceability of every batch of bottled wine back through all the processes and blends to the original grapes and even to the vineyard site from where they were harvested. This is a major problem for the producers of large volumes of table wine, where the final blend might well consist of numerous components from all parts of the country. The best way of achieving this by far is the installation of software that is specifically designed to track the movements and blending of batches of wine.

When large-volume wine is purchased for sale as a branded wine of a constant style, a technical specification should be agreed with the supplier that encompasses all the parameters of importance, including the appropriate tolerance or range. The pro-forma on the next page could form the basis of a specification agreement or merely a suitable format on which to record the results. Every shipment of wine should be analysed to prove conformation to this specification.

Laboratory analyses

The analyses described below are a selection that should be applied as a regular routine to all wines. The description of the method is meant only for interest and is not intended to be used as a laboratory methods book.

It is not essential for a winery to have its own laboratory. In fact it is becoming more common to have all analyses conducted by an accredited specialist wine laboratory. Analytical requirements are becoming more and more detailed, which means the purchase of advanced equipment that is very expensive. If this is added to the need for accreditation, it becomes obvious that the external laboratory is an essential part of the quality control of wine.

The range of analyses suggested below cover the natural components and parameters of the wine itself, followed by additives and contaminants.

TECHNICAL SPECIFICATION FOR WINE				
WINE: Medium White VdT			***Ref No***	05/219
ANALYSIS	**UNITS**	**Min.**	**Actual**	**Max.**
Alcohol	% vol	11.0	11.5	12.0
Total dry extract	g/l	24	28	32
Sugar-free extract	g/l	16	18	20
Residual sugars	g/l	8	10	12
Total acidity (tartaric)	g/l	7.0	7.5	8.0
Volatile acidity (acetic)	g/l	-	-	0.6
Free sulphur dioxide	mg/l	35	40	45
Total sulphur dioxide	mg/l	-	-	260
Ascorbic acid	mg/l			150
Sorbic acid	mg/l		nil	200
Potassium	mg/l	-	-	1000
Calcium	mg/l	-	-	100
Sodium	mg/l	-	-	30
Iron	mg/l	2	-	10
Copper	mg/l	-	-	0.5
Carbon dioxide	mg/l	600	800	1000
Dissolved oxygen	mg/l	nil	nil	0.3
Filterability index	min/100	-	-	30
Microorganisms	cols/ml	-	nil	100

•   *Density*

Density, *per se*, is of no particular value, but can yield useful information when used in combination with other measurements. The problem is that dissolved sugars increase the density, whereas alcohol decreases it.

Density is easily measured by using a hydrometer, which is a glass float rather like a fishing float, with a graduated stem. The lower the density, the more the hydrometer sinks into the liquid. The figure is read by aligning the eye along the surface of the liquid and reading off the value from the graduated scale.

It should be noted that there is still considerable confusion between density and specific gravity (SG). The two figures are different and are not interchangeable. SG is a comparison of the mass of a given volume of liquid with the mass of the same volume of pure water, and it thus has no units because it is a ratio. Density is the mass of a given volume of liquid at a given temperature and is usually, in Europe, expressed in units of grams per litre (g/l) at 20°C. This is the preferred expression nowadays.

There are many different units of density used around the world, the choice being dependent upon the tradition of the country:

Baumé (°Be) = specific gravity x 1000 (used in France)

Oechsle (°Oe) = °Be - 1000 (used in Germany)

For example, a wine of SG 1.085 = 1085°Be = 85°Oe

Brix bears no relationship with either Baumé or Oechsle, as it relates to the density of a solution of sucrose expressed as a percentage by weight. It is used simply because pocket refractometers are calibrated in this scale.

KMW (Klosterneuburger Mostwaage, Austria) = Brix = Babo (Italy) = Hungarian must weight.

A more accurate determination is made by using a gravity bottle, or pyknometer, which is a small bottle so designed that it will hold an exactly reproducible volume of liquid. This is then weighed on an accurate analytical balance that can read to 0.00001 gram.

Alternatively, it can be done automatically in a somewhat expensive instrument known as a densimeter that measures the vibration of a small glass tube into which the liquid is injected. The denser the liquid, the slower the vibrational period.

•   *Alcoholic strength*

Analysis of the alcoholic strength is important for two reasons. First, the price of table wine is linked in part to its alcoholic strength. A wine with an alcoholic strength of 11% vol will be more expensive than one at 10% vol, and it is important to know that the correct wine has been supplied. Second, all wines on sale in the EU must comply with the labelling regulations which allow a tolerance of only ±0.5% alcohol on the label declaration. It is therefore necessary to know that the label declaration is legal.

One could add a third reason for wanting to know the alcoholic strength of a wine, for many consumers imagine that the quality of a wine is linked to its strength and often look for the alcohol declaration before making the final decision to purchase. This is most unfortunate because, in reality, there is no connection between the two. The fact that Bordeaux Supérieur, for example, has a higher minimum alcoholic strength than simple Bordeaux does not necessarily mean that it is of higher quality. It is not the alcohol that enhances the quality; it is the extra concentration of the fruit elements in the juice that plays the important role; the higher alcohol is incidental.

The earliest known method of determining the strength of spirits was what was known as 'proving' the strength, hence the development of units of proof. The procedure was somewhat dramatic, involving wetting a small heap of gunpowder with the spirit and applying a lighted match. If the gunpowder ignited, the spirit was 'over proof'. If it would not burn then the spirit was 'under proof'. By experimenting with a range of strengths it was possible to calculate what became known as 'proof spirit', or 100° proof, and which actually contained about 57% alcohol by volume. A spirit of this strength would just allow the gunpowder to burn. The standard at which most spirits were sold was 70° proof, which is equivalent to the present day 40% alcohol. On the proof scale 100% ethanol registers 175°. This was all very confusing, particularly as the American proof was different, being precisely double the percentage by volume. Hence American spirits at

40% alcohol were labelled as 80° proof, giving the wrong impression that they were stronger than British spirits.

Thankfully, this confusion was swept away by Brussels when it declared that the unit of alcoholic strength would be the percentage by volume, abbreviated to % vol. It is worth noting that percentage by weight gives a considerably lower figure, due to the density of alcohol being much lower than that of water. The density of water at 20°C is 0.998g/ml, and that of pure ethanol is 0.789g/ml.

Ethanol is surprisingly difficult to analyse by simple chemical methods because it is not very reactive, so most methods depend on a physical characteristic, such as density or absorption of infra-red radiation, both of which lend themselves to automation.

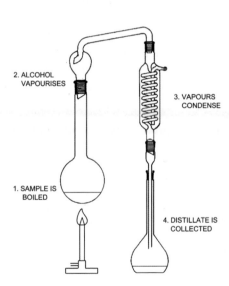

2. ALCOHOL VAPOURISES

3. VAPOURS CONDENSE

1. SAMPLE IS BOILED

4. DISTILLATE IS COLLECTED

Reference method for the analysis of alcohol

The classical method, known as the distillation method, makes use of the large difference in density between water and alcohol. It involves boiling a sample of the wine, passing the vapours through a condenser, and collecting the liquid that condenses. This process separates the alcohol from most of the other components in the wine that might interfere by affecting the density. An accurate determination of the alcoholic strength can be made by measuring the density of the distillate. Extensive tables are available giving the strength for every incremental change in density. This has been the official method for centuries, although it requires a great deal of expertise in the manipulation of the equipment if accurate results are to be obtained. Such is the esteem in which this method is held that not until 1997 was a modern instrumental method accepted as a reference method.

Laboratories in wine producing countries, where hundreds of samples are measured every day, are usually equipped with automatic instruments based on the measurement of the absorption of a beam of light in the near infra-red region (NIR) passing through the wine. A sample of wine is placed in the instrument, a button pressed, and the answer is printed out.

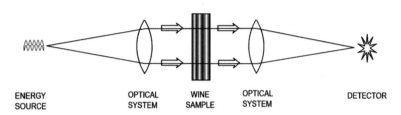

| ENERGY | OPTICAL | WINE | OPTICAL | DETECTOR |
| SOURCE | SYSTEM | SAMPLE | SYSTEM | |

Instrumental method for the analysis of alcohol using NIR absorption

The methanol and the higher alcohols are easily measured by modern techniques. Being volatile (easily evaporated by heat), they lend themselves to gas chromatography (GC), one of the most widely used of analytical instruments and easily adapted to analyse any volatile substance.

In this instrument, a small sample is injected into a long column containing an absorbent substance and through which is passing a current of an inert gas. The column is heated in an oven, which vaporises the alcohols. The different properties of the various alcohols cause them to pass along the column at different speeds, resulting in a separation at the far end of the column. By arranging for them to pass through a suitable detector, the individual alcohols can be recorded separately by what is known as a chromatogram.

All of the methods described above are aimed at the determination of the actual alcohol content of the wine, hence these figures are reported as 'actual alcohol'. If the wine contains some residual sugar, it would theoretically be possible to ferment it and produce more alcohol. This parameter is sometimes reported at 'potential alcohol', and the sum of these two figures is known as 'total alcohol'.

•   *Total dry extract (TDE)*

The name of this measurement is somewhat difficult to understand and requires explanation. It is defined as the total non-volatile substances remaining after evaporation of the wine to dryness under specified physical conditions, usually 100°C at atmospheric pressure. The water and alcohol will evaporate under these conditions, leaving the sugars, glycerol, non-volatile acids, mineral salts, polyphenols and other minor constituents.

In practice, dry extract is virtually never determined by this method, as it is too tedious and too lengthy. It can be ascertained with sufficient accuracy by using what is known as the Tabarié formula, which uses the figures for density and alcoholic strength to calculate the TDE, thus saving a considerable amount of the analyst's time.

What does this figure tell us? If the wine is dry, it gives some indication that the wine has not been subject to any falsification such as dilution with water. A dry white wine would be expected to lie within the range 16 – 20 grams/litre, which represents all of the non-volatile substances in the wine. The bigger the wine, the higher the figure.

If the wine contains residual sugars, then the total dry extract will give a useful approximation of the sugar content. For example, if it is a medium dry white wine with 12 grams/litre of sugar, then the TDE should lie between 28 and 32 grams/litre.

If the total residual sugars have been determined accurately, we can obtain another analytical parameter known as *sugar-free extract* by subtracting this figure from the TDE. This corresponds to the TDE expected from a dry wine, and similarly is a possible indication of fraudulent practices.

•   *Total acidity*

The name of this important parameter in the analysis of all wines came about because it is a measure of all the acids in the wine including the volatile acids (see below). An alternative name is titratable acidity because the analysis is carried out by a titration. The results are the same whatever the name, and even the commonly used abbreviation is the same for both.

Total acidity (TA) is determined by a simple titration, which will be familiar to all who studied chemistry at school. A pipette is used to measure a portion of wine into a conical flask, to which are added measured quantities of a standardised alkaline solution from a burette until the resultant mixture is neutral. This moment in the titration, known as the end-point, can be detected by using an 'indicator', which is a substance that changes colour when the mixture in the flask becomes neutral. Nowadays a pH meter (see below) is often used to detect the end-point giving greater accuracy, especially with red wines where a colour change may be difficult to see. The quantity of alkaline solution used is proportional to the total acidity of the wine.

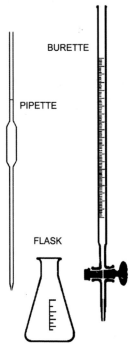

BURETTE

PIPETTE

FLASK

Essential laboratory glassware

The result has to be interpreted with care because it is not the measurement of a single acid but is the sum of all the different acids present in the wine. The only way to obtain a meaningful result is to express it as if there were only one acid in the wine, and different countries have different traditions which can cause confusion. The EU has adopted tartaric acid as the unit, which is logical since this is the most abundant acid. The French still persist in expressing the results as sulphuric acid, which seems a rather strange unit to use since it is hoped that their wines do not contain battery acid! The results yield a figure roughly half the size, conversion from sulphuric acid to tartaric acid requiring multiplication by 1.531.

The winemaker is interested in the total acidity at a very early stage, even prior to fermentation, for it is then that the first adjustment can be made. It is permissible, but less effective, to make a second adjustment after the fermentation. Thus a titration for total acidity would be carried out at both of these moments during the production of the wine.

There is one legal aspect of total acidity that should be observed, which is that for table wines on sale within the EU there should be a minimum of 4.5 g/litre, expressed as tartaric acid. (But wines as low as this will probably taste somewhat 'flabby' in any case.)

TA is one of those analyses that tend to be carried out because of its simplicity rather than its importance. Immediately prior to bottling there is little that can be done about total acidity. The important parameter is the taste: if the taste is correct, the wine is correct. There is, however, one very useful and unusual aspect of the measurement of total acidity that has been adopted by some bottling companies: it is a simple test for dilution and is used because of the ease of measurement. Total acidity is measured when the wine first arrives at the bottling plant and again in the first bottle from the bottling line. If the latter is lower than the former, this indicates that the wine must have been diluted by rinse water in the bottling line. The two results should be identical and bottling should not commence until this situation has been achieved.

• *pH*

For the non-scientist pH is a somewhat difficult concept to come to terms with, being an indication of acidity and yet not the same as total acidity and whose figures are the opposite way round. With TA, the more acid the wine the higher the figure, but with pH, the more acid the wine the lower the figure.

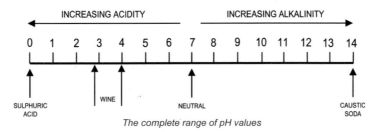

The complete range of pH values

The name pH comes from the French *pouvoir hydrogène*, meaning hydrogen power, and in scientific terms pH is defined as the negative of the log to base 10 of the concentration of hydrogen ions, or $pH = -\log_{10}[H^+]$, a somewhat difficult concept to comprehend. However, for the purpose of winemaking an understanding of the basic principles is all that is necessary. pH is expressed on a scale of 0 to 14, where 0 is very acidic (e.g. strong sulphuric acid), 14 is very alkaline (strong caustic soda) and 7 is neutral (as in pure water).

Most wines fall in the range 2.8 to 4.0. A highly acidic Muscadet, for example, might be pH 2.9, whereas a soft warm climate red wine such as Australian Merlot, might be pH 3.5. Although the range of pH found in wines is small, differences of 0.2 are important.

Its relationship to total acidity is not straightforward owing to what is known as the 'buffering' effect of the natural salts in the wine, which can cause a change in the pH without altering the total acidity. Thus it is possible to have two wines of the same total acidity, yet with different pH, or vice versa. However, pH is the more important parameter since this is the controlling function that affects colour, taste and keeping qualities, especially in relation to molecular sulphur dioxide (see p.173). It is probably best considered as an indication of the true acidity of the wine.

Measurement of pH is extremely simple by using a readily available and inexpensive piece of equipment known as a pH meter. All that is necessary is to dip the electrode into a sample of the wine and note the reading. Unfortunately the electrode gradually changes and must be calibrated every day before use. This is also a simple operation using readily purchased tablets which, when dissolved in water, yield a solution of accurately known pH.

• *Volatile acidity*

Volatile acid (VA) is so-called because it is the component of total acidity that is volatile, meaning that it can be separated from the other acids by boiling the wine. The major constituent is acetic acid, the product of oxidation of alcohol.

The classical method of determining volatile acidity is to boil off the volatile acids in a simple glass still and collect them as a distillate which is then titrated with alkali, as for total acidity. Because the volatile acidity is caused by a number of steam-volatile acids of different boiling points, the conditions of distillation must always be the same, including the design of the distillation apparatus, otherwise different results will be obtained. A second difficulty arises because of the formation of esters, which are not included in the titration as they are neutral and not acidic. Furthermore, if the acetic bacteria are active the VA will constantly be changing. It is therefore sometimes difficult to get agreement between laboratories on the results of this determination.

As with total acidity, the result is expressed as if it were one acid. In this case it is acetic acid, the acid of vinegar and the most abundant acid. The French, however, continue to use sulphuric acid as the standard. To convert from sulphuric acid to acetic acid the result must be multiplied by 1.225.

• *Residual sugars*

The measurement of residual sugars in the laboratory is traditionally done by a Fehling's titration, which involves a reaction with copper salts. Although somewhat complex in terms of chemistry, the titration is simple, and gives a sufficiently accurate measure of the residual sugars in wine. Fehling's Solution, which is widely used in food laboratories, was invented by H C von Fehling, professor of chemistry at the polytechnic school at Stuttgart from 1839 to 1882.

A modern high technology laboratory will probably use high performance liquid chromatography (HPLC), in which the individual sugars are separated by passing the wine through a column of absorptive material, in a similar way to the separation of alcohols by gas chromatography. The various sugars pass through the column at different speeds and thus appear at the detector at different times, producing peaks on a chromatogram that look somewhat similar to the results of gas chromatography as above. All of this can be automated, with the results stored in a computer, ready for instant retrieval when needed. This is a very elegant and advanced form of analysis which gives a complete picture of the residual sugars in the wine and is of particular use in experimental production.

It is not possible to obtain a figure for the sugar content by measuring the density, as with musts before fermentation, because alcohol is less dense than water and interferes with the density value. Similarly the refractive index reading cannot be used because the refractive index of alcohol is different from that of water.

• *Tartrate stability tests*

Although wine may have been refrigerated and should be tartrate stable, it is critically important to test the wine before bottling to ensure that stability has been achieved. The standard test for tartrate stability is to refrigerate a small sample at -4°C for three days and to examine the sample for traces of crystals. The presence of crystals at the end of the test indicates that the wine is unstable and that it should be treated further. If crystals are absent, it is unfortunately no guarantee that the wine is stable because colloids might be present which are preventing the deposition of crystals.

Tests that are more elaborate have been developed, one of which is based on the measurement of the concentration of all the substances in the wine that affect tartrate stability, viz. potassium, calcium, tartaric acid, alcohol and pH. After these have been measured, the results are entered into a formula which indicates whether the wine is stable or unstable.

Another test, which seems to be the most effective and practical, makes use of the change in electrical conductivity which takes place when there is a change in the concentration of dissolved salts in the wine: the stronger the solution, the greater the conductivity. In this test the wine is stirred at zero degrees with finely divided potassium bitartrate crystals and the change in electrical conductivity is measured. A rise in conductivity indicates that crystals are dissolving, showing that the wine is not saturated with tartrates and therefore should be stable. If the conductivity goes down, this indicates that tartrates are crystallising out, showing the wine to be supersaturated and thus unstable. Although better than the more simple tests, the most reliable results are produced with wines that are well known to the technician because each wine has its own characteristics. In other words, this test works best for producers, who know their wines well, rather than bottlers, who have to handle large numbers of wine, all of which are different and have somewhat unknown properties.

Nevertheless, all of these tests fail from time to time because of the complex structure of wine with its many colloidal substances, all of which interfere with normal crystallisation processes. In view of the unreliability of this entire subject, one might think that it would be easier to teach consumers to accept tartrate crystals!

• *Protein stability tests*

Unstable proteins that cause potential problems of haze, cloudiness or deposit are easy to detect by submitting a small sample to abuse by heating it to 80°C and cooling it to room temperature. If it remains clear and bright, it is stable. If it produces a haze, or deposit, the wine needs fining. This test is an excellent example of an empirical quality control test, where extreme conditions are simulated in the laboratory and whose results indicate the treatment that is necessary.

It is also possible to detect unstable proteins by adding a solution of phosphomolybdic acid, widely available as the proprietary Bentotest reagent, that produces an immediate coagulation of the proteins, resulting in a visible haze. This can be a useful test but tends to be over-sensitive, resulting in unnecessary treatment of the wine. As with the tartrate stability test, interpretation of the results is at its most reliable when used by producers, to whom the characteristic behaviour of the wine is well known.

Permitted additives

The addition of any permitted additive must always be accompanied by a subsequent analysis, to ensure that the correct level has been added. Despite the best controls in the most modern of wineries, it is not unknown for additions to be made twice, or for two operatives to assume that the other has made the addition, when neither has; hence the importance of keeping good records.

• *Sulphur dioxide*

Checking for the correct level of **free sulphur dioxide** is the most frequently performed test in the quality control regime because the protective action of sulphur dioxide is self-destructive. When oxygen dissolves in wine, the sulphur dioxide should destroy the free oxygen by reacting with it before it has a chance to oxidise any of the wine components. During this reaction the sulphur dioxide is itself destroyed by being converted to sulphuric acid (in minute quantity!). Further additions have to be made at every stage in the wine making and handling process to keep the sulphur dioxide at the correct level.

Fortunately, the standard test is simple to perform and can be carried out on a table in a corner of the winery. It is possible to purchase a small kit of glassware and solutions from many laboratory suppliers to enable a titration to be performed in which a standard solution of potassium iodate, or iodine, is used to oxidise the sulphur dioxide, giving a direct reading of the free sulphur dioxide content of the wine.

The disadvantage of this simple method is that iodine is a reactive chemical and will oxidise other substances in the wine as well as sulphur dioxide. This is a particular drawback when the wine contains ascorbic acid, since iodine reacts with it as if it were sulphur dioxide. For results that are more accurate, a so-called 'blank' titration has to be carried out by first adding acetaldehyde to a second sample of the wine. This binds up the entire free sulphur dioxide before the iodine titration is carried out. The result is the totality of all the other substances in the wine that will react with sulphur dioxide, and this 'blank' figure has to be subtracted from the first titration before calculating the result.

The only way of obtaining a truly accurate result is by using the official EU reference method, variously called the peroxide method, or the aspiration method, aspiration meaning drawing something through by suction.

In this case a current of air is slowly bubbled through a sample of the wine in a special aspiration apparatus, as shown in the diagram below, carrying with it the free sulphur dioxide and leaving behind the non-volatile interfering substances.

The current of air bearing the sulphur dioxide is then passed through a solution of hydrogen peroxide, which oxidises the sulphur dioxide to sulphuric acid. The sulphuric acid can then be titrated with a standard solution of sodium hydroxide.

The same kit can be used for measuring the **total sulphur dioxide** by first treating the wine sample with an alkaline solution, which has the effect of releasing the bound sulphur dioxide and converting it to the free form. The total sulphur dioxide can be obtained by performing a titration after this treatment. As it is the total sulphur dioxide that is controlled by EU regulation, it is important that this level is known every time before any additions are made to ensure that the wine is kept within the legal limits.

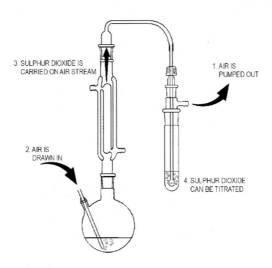

3. SULPHUR DIOXIDE IS CARRIED ON AIR STREAM

1. AIR IS PUMPED OUT

2. AIR IS DRAWN IN

4. SULPHUR DIOXIDE CAN BE TITRATED

The classical method of analysis of sulphur dioxide

The same caveat applies to the measurement of total sulphur dioxide as to the measurement of free sulphur dioxide: the simple iodine titration is prone to give falsely high results because of interference by other substances in the wine. With total sulphur dioxide subject to legal limits, in any case of litigation the analysis must be performed by the aspiration method.

• *Other additives*

Simple analytical methods are available for ascorbic acid and sorbic acid. As there are legal limits for both of these substances, analysis should always take place after any additions have been made to ensure that the limit has not been exceeded.

Methods for the analysis of metatartaric acid and citric acid are more complex and are used only in those special cases where it is considered necessary for purposes of verification of legality, both of these additives also having legal limits imposed by EU regulations.

Contaminants

• *Dissolved oxygen*

Oxygen will dissolve very readily in wine that is in contact with air, and once dissolved, the oxygen begins its steady task of destroying the wine. Unfortunately, the sulphur dioxide that is also present in the wine does not react immediately with the oxygen, but co-exists with it for sufficient time to cause damage. Hence, the only safe practice to observe is to prevent any oxygen from dissolving, and to monitor constantly the level of dissolved oxygen in the wine.

At normal room temperature, one litre of wine can dissolve about eight milligrams of oxygen before becoming saturated. In a tank of wine, filled to the brim and properly sealed, this will gradually fall below one milligram per litre after about ten days, due to the scavenging effect of the sulphur dioxide. More oxygen dissolves every time the wine is moved and this should be measured and monitored so that techniques can be improved to minimise this.

The dissolution of oxygen increases at low temperatures because oxygen becomes more soluble. Wine at -5°C can dissolve almost twice as much oxygen as at 20°C. Therefore wine that has been chilled must be handled with even greater care than when it is at room temperature.

Dissolved oxygen is one of those analyses that can be performed very simply by using a specifically designed meter. In common with many of these electronic devices, the operation of such a meter is simple but false results can easily be produced. First, the electrode tends to go out of calibration rather quickly, so requires constant checking. Second, because we are surrounded by an atmosphere containing 20% oxygen, care has to be taken to prevent high results.

The only truly accurate way of measurement is by using a meter that pumps the sample through a closed cell, thus eliminating any possibility of a false result due to contamination by atmospheric oxygen during the analysis. Frequent calibration is still a necessity.

• *Iron and copper*

Although both of these elements occur naturally in all grape juice and wine an excessive level indicates that contamination has occurred. An iron concentration above the natural level of around 10 mg/litre indicates the use of old galvanised buckets or ancient presses with bare iron components. Copper levels are normally much lower, being less than 0.2 mg/litre. An elevated copper content is usually caused by bronze fittings or pumps in the winery, or late spraying of copper fungicides in the vineyard. Both elements are undesirable in excess because they are the cause of both casse and oxidation.

A small quantity of copper is desirable in wine, acting as a cleansing agent and removing traces of hydrogen sulphide (H_2S), the cause of the 'bad egg' smell that sometimes becomes apparent when sulphur dioxide becomes reduced. This phenomenon occurs more frequently now that copper and bronze equipment has been removed from wineries and everything is made of inert stainless steel. Winemakers sometimes have to add a small quantity of copper sulphate to precipitate the sulphide as the very insoluble copper sulphide.

It is interesting to note that in very old-fashioned wineries, where even the pipework is made of copper, the wine is not heavily contaminated with copper. The inner surfaces become covered with black copper sulphide which forms an inert coating.

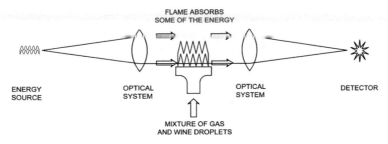

Analysis of iron and copper by atomic absorption spectrophotometry

In the days of classical analysis, determination of iron and copper was tedious and difficult. Nowadays it is made simple with modern instrumental equipment such as an atomic absorption spectrometer (AA). A beam of light from a special lamp with electrodes made from the element being determined is passed through a gas flame. The wine is sprayed into the flame in the form of a fine mist, and the change in the intensity of the beam is measured. This gives a very specific and accurate way of checking the concentration of different metals in the wine; all that is needed is a special lamp for each element.

• **Sodium**

Wines are naturally rich in potassium and very low in sodium, normal levels being 1000 mg/litre and 20 mg/litre respectively. Ion exchange alters this ratio considerably, a change that is particularly noticeable in the sodium concentration, which can be raised by a factor of ten. Thus, measurement of the sodium content of a wine is a simple test for ion exchange. This is easily carried out with equipment such as a flame photometer, which measures the emission of light at the characteristic sodium wavelength, the yellow light with which we are all familiar in the form of street lighting.

Advanced methods of analysis

The analyses described above are the classical range that have been used for many years in wine laboratories. Nowadays, DNA analysis can be used to confirm the origin and vintage of a wine by reference to databases that have been developed over the last decade.

Gas chromatograpy (GC) and high performance liquid chromatography (HPLC) are not new to wine laboratories, but their value has been greatly increased by the addition of mass spectrometry (MS). Chromatography is a technique that is used to separate the components of a mixture by passing it through a tube containing an absorbent material. The different components pass though the tube at different speeds, hence they are separated. This has long been a useful technique, but identification has not been easy until the advent of MS. This instrument can be attached to the outlet of a GC and will identify the individual components by breaking them down into their specific ions. And an even more accurate result can be obtained by placing a second MS at the outlet of the first, this method being known as GC/MS/MS.

So sophisticated have modern laboratories become that, despite the cost, multiple GC/MS/MS set-ups are maintained, each one for a specific task because it eliminates time taken in changing the parameters for different analyses.

Microbiological analysis

The yeasts and bacteria that play an important role in the production of wine are unwelcome entities when the time comes to put the wine into a bottle. Their presence is of no great importance in traditionally made dry wines because the bacteria cannot survive in the absence of oxygen and the yeasts will quickly die, if not already dead, due to lack of nutrients. These wines are naturally stable.

The situation with wines containing residual sugar is, however, quite different. If these wines still contain viable yeast cells when bottled, a second fermentation will start in the bottle with disastrous results. Hence the importance not only of a good technique of aseptic bottling but also a reliable procedure for checking for the absence of microorganisms.

Photomicrograph of a colony of Saccharomyces cerevisiae grown in three days from a single cell

The picture above shows the colony of yeast that has developed in only three days from a single cell of *Saccharomyces cerevisiae,* hence the need for high standards of hygiene in all wineries.

A remarkably simple and elegant test has been developed which is widely used and requires no expensive equipment, neither microscope nor a sterile room. All that is necessary is a specially designed funnel which will hold a small circular piece of membrane filter in the base, a flask into which the funnel fits, and a vacuum pump.

Before commencing the test the funnel is sterilised either by steaming or by covering with alcohol and burning it off. The membrane filter is purchased already sterilised and is fitted to the funnel, and the funnel placed in the flask. The wine sample is poured into the funnel, the vacuum pump turned on and the wine is sucked through the membrane, leaving any microorganisms stranded on the surface of the membrane.

The membrane is then removed from the funnel and placed on the surface of a nutrient medium in a sterile petri dish. The nutrient diffuses into the pores of the membrane and nourishes any living cells that might be present, which reproduce rapidly. After three days each cell grows into such a large colony that it can be seen with the naked eye. All that is necessary is to examine the membrane and count any colonies that might be visible. A successful bottling is indicated by a membrane that is as clean as when it was first removed from its packet.

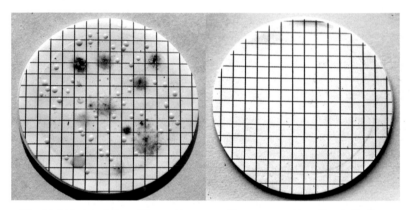

Microbiological swab from dirty machinery   *The same machine after sterilisation*

The photographs above are the results of tests from a filling machine before and after sterilisation. In the picture on the left the creamy coloured compact spots are yeast colonies, each of which started as a single yeast cell which has multiplied thousands of times to become a colony visible to the naked eye. The flatter and broader growth near the seven o'clock position is a bacterial colony and the fluffy areas are colonies of mould, indicating a filler that is highly contaminated with various micro-organisms and is in need of sterilisation. The clean result above is the norm for a well-managed modern bottling line.

Chapter 20

WINE FAULTS

Wine kept for two or three years develops great poison.

Chinese 14th century

Old books on the technical aspects of wine list a range of troubles that are virtually never encountered nowadays, many of which were of microbiological origin and have been eliminated by improvements in filtration and sterile techniques. Better control of antioxidants has reduced the number of oxidised bottles on the shelf, and the improved quality of packaging materials has almost eliminated foreign body complaints.

The following problems remain and will be difficult to eliminate totally, and consumers should not be afraid to take bottles back to the retailer. Although all retailers are very willing to refund the money or replace the bottle, most take little interest in recording the proper details of the fault, simply registering the returned bottle as 'Quality complaint'. This is most unhelpful as the producer of the wine has no facts on which to base any investigation, with no chance of continuous improvement or of reducing the incidence of faulty bottles. It has also resulted in the cork being wrongly blamed when it has been entirely innocent. This situation has come about because many people, including highly qualified members of the wine trade, have no idea how to recognise faults in wine.

Oxidation

This is probably the most common cause of complaint nowadays because oxidation can occur in any wine at any stage of its life. It occurs in wines when oxygen from the atmosphere gets into the package via a poor closure or simply by migration through the material of the package itself. They lose the protection of antioxidants, both natural and added, and the oxygen attacks the wine, causing the breakdown of the fruit components.

The first indication of an oxidised wine is a loss of its attractive colour:

- White wines that should be a clear lemon yellow with green tints become a dull brown straw.

- Rosé wines that started life as vibrant pinks become a pale orange-pink, or even a brown pink. (Caution is needed here, as some rosé wines are an orange-pink when first produced.)

- Red wines lose the deep purple red or ruby colour and become paler, with an orange or brown rim.

The bouquet of the wine becomes tired, losing its freshness and smelling of caramel or even 'meaty' (like beef extract). The palate confirms the impressions of the nose. Ultimately, the wine takes on the character of Madeira, which is a deliberately oxidised wine. Hence, oxidised wines used to be described as 'maderised'.

When a wine is packed it should have been brought into the correct condition by first carrying out a complete analysis, followed by any necessary treatments or additions, especially the adjustment of the free sulphur dioxide level. However, if the wine has been allowed to absorb quantities of oxygen from the atmosphere because of careless handling or storage, the sulphur dioxide in the wine will quickly be destroyed after packing, leaving the wine prone to early deterioration.

Those materials with the greatest oxygen barrier, such as glass bottles and metal caps, will confer on the wine the maximum shelf life. Packages such as plastic bottles and bag-in-boxes which, despite the best efforts of technologists, do not provide a total barrier to oxygen, result in a reduced shelf life. Stock control of these items is of paramount importance if stale wine is to be avoided.

Reductive taint

This is a comparatively recent phenomenon resulting from the use of screwcaps or other closures that prevent the ingress of oxygen (see p.206). It is caused by the reduction of sulphur dioxide to hydrogen sulphide, and imparts a dirty odour of drains or bad eggs to the wine. At very low levels, the wine merely smells unattractive or slightly metallic. The cure is treatment with a little copper, added as copper sulphate. In the tasting room it is easily confirmed as reductive taint by adding a copper coin the the tasting glass and swirling gently. After a few minutes the wine will smell fresher and cleaner, if the cause really is reduction. The copper ions dissolved from the coin precipitate the sulphide ions that are the cause of the problem. If the wine does not become fresher, the problem lies elsewhere.

Beyond shelf-life

Contrary to the belief of many consumers, wine does not have an infinite life in bottle; neither do all wines improve with keeping. In fact, more and more wines are made for immediate consumption and actually deteriorate with time. A glance at the back label of wines on a supermarket shelf will soon confirm this, with the frequent recommendation that 'This wine should be consumed within six months of purchase'. This is the popular modern style of wine for which many consumers are searching, but does not have sufficient structure for long life. Wines that improve with keeping must have more concentration, more body, and especially higher levels of the polyphenols, the natural preservatives.

Although oxidation and shelf-life are inextricably linked, it is important to differentiate between oxidation and natural decomposition. Wines that have reached the end of their natural life are not necessarily oxidised. The fruit components simply break down, leaving the wine dull and lacking in any character, and this can happen in the absence of oxygen. A good example of this phenomenon is fino sherry: this wine should be drunk within six months of leaving the solera, not because of oxidation, but because it is intrinsically unstable and relies on the protection of the flor yeast and the frequent topping-up with younger, fresher wine.

The 'surface effect' (see p.212) also influences the shelf life of wines in all forms of package. The longest shelf life occurs in a large bottle made of glass, such as a magnum (1.5 litres), or even better, in a nebuchadnezzar that holds 16 litres. The worst conditions are to be found in a small bottle such as the standard airline size of 18.7 cl, especially if it is made of PET. Such bottles are in use, and are ideal for certain circumstances such as in-flight service, but the short life of only some three to six months has to be taken into account when planning for stock holding.

It might seem strange that there is no provision for a "best before" date on wine labels. There is a general requirement in Europe that all foodstuff with a shelf life of less than 72 weeks should carry such an indication. However, wine is exempt from this legislation, presumably on the assumption that all wine has a shelf life exceeding 72 weeks. This, of course, is incorrect in the case of small PET bottles and bag-in-box. Nevertheless, this anomaly remains and a "best before" date does not have to be displayed. The ideal is to give a recommendation for usage before a set period after purchase, as above (see also p.199). Stock control is also of paramount importance because it determines the age of the wine before purchase by the consumer.

Tartrate crystals

Tartrate crystals in a bottle are regarded as a fault by most consumers, despite any arguments to the contrary. This deposit is not tartaric acid, as is commonly thought, but is either calcium tartrate or, more likely, potassium bitartrate (cream of tartar), originating in the grape. This is the result of either a poorly conducted stabilisation process or the initial presence of protective colloids that have subsequently denatured. They may be intentional, in that the winemaker preferred to bottle the wine untreated.

Potassium bitartrate crystals

Two understandable fears that are generated in the minds of consumers are that the wine contains broken glass or added sugar, although this latter observation is irrational because sugar would dissolve in the wine.

Tartrate crystals in a bottle of wine are probably the biggest cause of complaint in the wine trade. The best that can be done to prevent this phenomenon is to fine the wine thoroughly, testing for the complete removal of colloids, and then to submit it to treatment by the contact process, testing for tartrate stability by one of the modern methods. Unfortunately, crystals sometimes still come down, despite the best efforts of winemakers and bottlers. Perhaps an explanatory back label would make the crystals acceptable, especially if imaginatively worded, as in one old German label that read, "This wine contains Wine Diamonds, which are an entirely natural deposit"!

A great deal of expense, energy and effort have gone into the prevention of these deposits, adding a significant amount to the cost of the wine, with also a possible loss of quality. All that is necessary is to decant the wine slowly, when the crystals will remain in the bottle. If they get into the glass they will probably remain there, stuck by surface tension and even if they get into the mouth, they are completely harmless, merely tasting slightly bitter. If only the consumer could be educated to understand what tartrate crystals are and why they are there, life would be so much simpler for the winemaker. It is ironic that the winemaker goes to a great deal of trouble and expense to remove these harmless tartrates and then sells them to the baking industry as cream of tartar.

The only way of dealing with stocks of wine that are showing crystal deposits is to open the bottles, tip the wine back into a vat, submit the wine to refrigeration and re-bottle. It is quite possible that the refrigeration treatment is unnecessary, all the crystals having been deposited in the original bottles, but it is safer to err on the side of caution. It is utterly devastating to a business to commit the same offence twice running.

Tartrate crystals can be prevented by cold stabilisation, contact process, electrodyalisis, ion exchange, or by the addition of metatartaric acid, cellulose gums or mannoprotein (see chapter 14).

Foreign bodies

It is just possible that what might be thought to be a tartrate crystal is actually a fragment of glass. If the nozzles on the filling machine become bent because of badly formed bottle necks or poor maintenance, small pieces of glass can be chipped off the rim of the bottle. Or possibly, after a bottle breakage, pieces of glass might find their way into other open bottles on the conveyor.

Other foreign bodies that can get into the wine include flying insects, human hair, or even parts of the filling machine that have come loose – it is not unknown for a bottle to contain an entire filling nozzle!

The good operators are well aware of this danger and will have it highlighted in their list of critical control points in their HACCP procedure.

Musty taint

Wine tainted by a cork is commonly called 'corked' or 'corky'. This is an unmistakable musty smell resembling fungi or autumnal woodlands, and renders the wine undrinkable although still harmless. Such bottles should be sent back or returned to the retailer. Care should be taken before making a final judgment, however, because a somewhat similar earthy note is a natural characteristic of some grape varieties. Furthermore, the untrained nose finds it difficult to distinguish between musty taint, oxidation and other out-of-condition wines, frequently blaming the cork when it is entirely innocent.

The cork industry has, with some justification, expressed annoyance at the way in which cork is blamed for every instance of musty taint. The two essential ingredients for the generation of 2,4,6-trichloranisole (TCA) are phenols and chlorine. It is quite possible for TCA to be generated by the use of phenolic wood preservatives and hypochlorite sterilants. Many wineries have totally banned the use of all chlorinated sterilants in favour of peracetic acid and ozone.

It should be particularly noted that wine containing small pieces of cork as a result of a brittle cork breaking up is *not* corky. It is not even faulty.

The remedy here is simple: remove the pieces with a finger or a spoon and carry on drinking!

Incidentally, there is no need to waste wine that has been contaminated with TCA: it can be used in cooking because the TCA is steam volatile and boils off during the cooking process.

Volatile acidity

Wine with excess volatile acidity is the result of careless wine handling techniques, with bacteria running rampant, a shortage of free sulphur dioxide and an ample supply of dissolved oxygen, all of which are avoidable situations. The bacteria convert alcohol to acetic acid (the acid of vinegar) which then reacts with further alcohol to produce the ester known as ethyl acetate. The wine becomes undrinkable, smelling of nail varnish or cellulose paint (both of which are based upon ethyl acetate) and tasting of vinegar (acetic acid).

The only use for such a bottle is to consign it to the vinegar pot and encourage it by aeration to continue its evolution to wine vinegar.

(For further information on volatile acidity, see p.138.)

Second fermentation

Wines made by fermentation to dryness, and containing no fermentable sugars, are stable towards further fermentation and can safely be bottled by traditional methods without the risk of a second fermentation occurring. However, this is not the case with many modern wines that contain fermentable sugars, however small in quantity. These wines are susceptible to re-fermentation by stray yeasts and must be bottled aseptically.

When a yeast starts a fermentation in the bottle the effects are obvious: the wine goes cloudy and the cork starts to rise due to the pressure of the carbon dioxide that is being produced. Bag-in-box packages can be even more dramatic, blowing up like footballs and eventually bursting, with the resultant chaos of collapsing pallets and aromatic warehouses!

The only treatment for wines that are undergoing a second fermentation is to disgorge them and filter them quickly to remove the yeast. After analysis and correction of the sulphur dioxide content they can be re-bottled, provided the style of the wine has not been altered. The procedures in the bottling plant should then be closely examined to determine the cause of the problem, and corrective action implemented. Adjustment of the sulphur dioxide level is particularly important in these circumstances because the yeast activity destroys all the existing sulphur dioxide.

The incidence of second fermentation has diminished greatly in recent years. With improved equipment and better knowledge this fault should be extremely rare. Its presence indicates bad hygiene or ignored procedures.

Iron casse

Excess iron dissolved in wine can result in a white precipitate or deposit in the bottle formed by a reaction between iron and the phosphates in the wine. This is traditionally known as iron casse, from the French word 'casse', which has the same meaning as the English word 'precipitate', something which has been thrown out of solution. The prevention of this fault is either by blue fining to reduce the iron level (see p.153), or by the addition of citric acid which prevents the formation of the precipitate.

The precipitate is harmless, as is the wine, which can be decanted off the solid matter and drunk normally.

Copper casse

Copper casse is easily recognisable as a brown haze in the wine, with the mysterious property of disappearing when the bottle is opened. This is because copper casse is a complex of cuprous ions and proteins which can only form in anaerobic conditions such as are found inside an unopened bottle. When the bottle is opened and the wine becomes aerated, the cuprous ions are oxidised to cupric ions which breaks the complex and the haze disappears.

This fault is easily prevented by analysis of the copper content as part of the preparation of the wine for bottling. If a high copper level is indicated, it should receive a treatment of blue fining (see p. 153), followed by another copper analysis to ensure that the wine is satisfactory for bottling.

Wine with copper casse should be regarded with more caution than wine with iron casse because copper is toxic at high concentrations, although it is an essential trace element in the diet.

Mousiness

Sometimes, but rarely in these days of improved bottling techniques, a wine can develop a smell of mouse droppings, often more apparent on the aftertaste than on the nose. This is due to an infection of a lactic acid bacterium. These organisms are easily controlled by sulphur dioxide and can be removed from the wine by membrane filtration prior to bottling. A mousy smell is a sign of poor hygiene, and once it has affected the wine there is little that can be done in the way of rescue.

Brett

Probably the most controversial of all microbiological phenomena, Brett is sometimes regarded as beneficial, sometimes as a fault. The smell is variously described as 'animal', 'farmyards' or 'horsey'. It is caused by various species of the *Brettanomyces* yeast, but particularly *Brettanomyces bruxellensis*. Wines from Chateau Beaucastel famously have high levels of Brett - but they are rightly regarded as superb wines.

It seems to be a matter of controlling the level during winemaking: a certain level adds complexity, where too much becomes a fault. As it is widely distributed, and its tolerance towards sulphur dioxide is variable, a few large doses of SO_2 seems to be the best method of control. Personal preference undoubtedly plays a large part in the appreciation of the Bretty aromas.

The various smells are due to the presence of three compounds that are produced by the *Brettanomyces* yeast:

- 4-ethylphenol - farmyard, antiseptic plasters (unpleasant);
- 4-ethylguaiacol - bacon, spice, cloves, smoky (attractive);
- isovaleric acid - cheese, rancidity, sweaty saddle (unpleasant).

Geranium taint

Another result of bacterial contamination can be the development of a powerful smell of geraniums (pelargoniums, to be precise). This is caused when a certain strain of lactic acid bacteria infects a wine containing sorbic acid which the bacteria metabolise, producing a substance known as 2-ethoxyhexa-3,5-diene. As with mousiness, wines that have developed this fault are fit only for destruction. Likewise the root cause is poor winery hygiene, but this problem is avoidable simply by not using sorbic acid as an additive. There should be no reason for its use because good standards of filtration and sterile techniques render it obsolescent. (See p.175 for more information regarding the use of sorbic acid.)

Chapter 21

THE TASTE TEST

He who loves not wine, women and song remains a fool his whole life long.

Martin Luther, 1777

There are many books devoted to the art of tasting, so there is no intention of going into an extended treatise on the technique of oral assessment. Conversely, as this is a book on the technology of wine, a few words on the technology of tasting might be in order.

Preparations for tasting

Temperature

In a commercial environment, when time is limited, most wines are tasted at room temperature. This is probably the most cruel way of assessing any wine, but especially whites and pinks as all of their faults are fully exposed. On the other hand, this is the best way of selecting the good from the bad, nothing being concealed.

When a wine is being savoured for its true purpose, then temperature is critically important. Unfortunately, many people take a little knowledge too far, knowing that white wines generally should be drunk cool and reds at room temperature. The result is that white and pink wines are chilled to mouth-numbing temperatures and reds are served positively tepid.

A white or pink wine served too cold loses its aroma, as the compounds responsible for the nose become less volatile and remain firmly anchored in the liquid. On the palate, the extreme cold numbs the taste buds, and the volatiles have little chance of reaching the olfactory organ in the back of the nose. These wines should never be served at a temperature that causes condensation on the bottle or glass, but should be around 8 to 10°C.

Over-heated red wines can be completely ruined by the excessive amount of alcohol that volatilises into the nose, and on the palate they taste flabby and unbalanced. This is a frequent problem in our modern centrally-(over)heated rooms. Don't be afraid to ask for an ice-bucket for your red wine in a restaurant, even if your request is greeted with a bemused look from the sommelier! When a red wine hits the palate, it should feel slightly cool, and should be served between 15°C for light red wines and no more than 18°C for heavier styles.

Decanting

A great deal of nonsense has been written about this very emotive subject. There can be no doubt at all that a whack of oxygen just prior to drinking (or tasting) a wine that has been shut in a bottle for some months or years yields quite an improvement. After all, the wine has been dosed with sulphur dioxide as a preservative prior to bottling and it has been in a reductive state during its sojourn in its container. It is desperate for a whiff of oxygen, which enables the fruit to develop and the wine to blossom.

In order to achieve this, it is useless simply to remove the cork and leave the bottle standing. The surface area in contact with the atmosphere is roughly four square centimetres, the wine is static, and oxygen dissolves only slowly. Very little will happen, unless the wine is decanted.

There are two reasons for decanting: to separate a mature wine from its sediment, or to aerate a young wine. In the first case, the wine should be decanted slowly, just before drinking, into a slim, upright decanter. If aeration is required, then the wine should be tipped quite roughly into a broad, shallow carafe, with plenty of swirling to dissolve the oxygen, and then left for an hour or two before drinking. The improvement can sometimes be quite amazing – and this applies to young wines of all colours. Even wines of considerable age sometimes develop after oxygenating, despite the danger of them breaking up.

Tasting (or drinking) glasses

The size and shape of the glass has a considerable effect on the way wine is appreciated. This effect is particularly pronounced on the nose.

The ISO tasting glass

The ISO tasting glass (ISO 3591 : 1977) is a good design for general use in a tasting room. It has a total capacity of 210 ml and holds 50 ml of wine when filled to the correct level, which is when the surface of the wine is at the level of the largest diameter of the glass. This gives a volume of 160 ml for the aromas, and plenty of space for swirling.

It cannot be emphasised too strongly that all students should equip themselves with a set of such glasses. The use of a variety of shapes and sizes will put them at a considerable disadvantage when it comes to tasting examinations. All examination bodies use ISO glasses for making their own assessments, and these are the benchmarks used for the marking of papers.

It has to be said that many tasting professionals feel that this glass is too small and does not give the wine a fair chance to show its best. There is a tremendous range of glasses available nowadays produced by several glass-blowing companies, notable amongst which is Riedel. The style labelled 'Chianti' has been approved by the Institute of Masters of Wine for use at its student training seminars.

Although there can be no doubt that the shape of the glass does have an influence on the perceived characteristics of the wine, it is not strictly necessary to own a shed-full of vinous glassware. In general, the bigger the wine, the bigger the glass, not so that it will hold more wine, but so that the more voluptuous bouquets can have adequate space in which to show off their qualities. The main considerations are that the glass should be in-curving towards the top and made of thin glass of high crystal quality so that it is completely colourless. It is unfortunate that the fashionable shape of the last century was the conical glass made of thick lead crystal covered with cut patterns. Beautiful though these glasses are, they do a terrible disservice to the wine and are best relegated to the china cabinet.

Styles of tasting

Tasting in front of the label

When learning about the different characteristics of the myriad wines of the world, the only way is to do it in front of the label, and preferably with a more experienced taster to give guidance. Constant practice is required (oh, what a hardship!) in order to train the palate to recognise the various nuances. It often takes a considerable length of time to become proficient and it is sometimes depressing to feel the lack of progress. But persistence pays dividends.

Comparative tasting

This style of tasting is sometimes called 'partially specified' because this is what it is: some common factor is known about the wines, but not sufficient to identify them specifically. It might be that they are of the same grape variety, or from the same vintage, or from the same region, district or even property.

This is a very useful approach to learning how to taste analytically, because the common factor gives a number of clues about the wine that should help to identify it. But, and this in an important but, for those taking examinations it must be remembered from the beginning that the objective of blind tasting is *not* to identify the wine, but to be able to give a good description of it and to assess its quality.

Blind tasting

The ultimate test is to be able to taste a wine in a glass and to tell everybody what it is, down to vintage and vineyard. Or is it? This might be a good party game, but it is not the aim of professionals in the wine trade.

'Unspecified tasting', as it is sometimes called, requires a quite different technique from tasting in front of the label. Firstly, it is necessary to empty one's mind, concentrate entirely on the senses, and then taste without any preconceptions. This can be quite difficult.

Having received the first sensations, one must progress to the end of the exercise, again with an open mind. Only when all observations have been completed should an attempt be made to assess the results. One of the greatest dangers in blind tasting is to come to a conclusion too quickly and then to make all subsequent observations fit the initial decision.

It is important to realise that the outcome of this tasting is not necessarily to decide the origin of the wine. This can be a 'fun' thing to do at a party, but can also be somewhat embarrassing, so should be handled with care. Of far greater importance is the assessment of the wine itself: its intrinsic quality, its expected price bracket, its age and development. The ability to assess these parameters correctly is key to passing the practical sessions in examinations.

Many students express fear and worry when sitting tasting examinations. This is needless if the above paragraph is considered carefully. Success in these examinations is not about getting the wines 'right' or 'wrong': it is the ability to assess the important parameters of each wine correctly, which is easy if one loves wine!

Constant practice is necessary, as with other forms of tasting exercise, especially in the period leading up to an examination. The only way of achieving this is to find somebody who can select and set up wines for tasting blind, and then to be very disciplined and approach each one in a proper student manner, writing a formal tasting note.

Writing a tasting note

There are two distinct reasons for writing notes about the taste of wines:

- To record the taste of a wine for future reference;

- To communicate the experience to another person.

It is important in both of these cases that simple descriptive words are used so that re-call and communication are meaningful.

W I N E   T A S T I N G
A Systematic Approach

A P P E A R A N C E

Clarity   : bright - clear - dull - hazy - cloudy

Colour   : white: green - lemon - straw - gold - amber
                   rose : pink - salmon - orange - onion skin
                   red   : purple - ruby - garnet - mahogany - tawny

Intensity   : white: water-white - pale - medium - deep
                   rose : pale - medium - deep
                   red   : pale - medium - deep - opaque

Other observations   : legs, bubbles, rim vs core, deposits

N O S E

Condition   : clean - unclean

Intensity   : weak - medium - pronounced

Development   : raw - grape aromas - mature bouquet - tired -
                                                           oxidised
Ripeness   : green - ripe - over-ripe - noble rot

Fruit character: fruity, floral, vegetal, spicy, woods, smoky, estery,
                 animal, fermentation aromas, faulty, (complexity)

P A L A T E

Sweetness   : dry - off-dry - medium dry - medium sweet -
                   sweet - luscious

Acidity   : flabby - balanced - crisp - acidic

Tannin   : astringent - hard - balanced - soft

Body   : thin - light - medium - full - heavy

Fruit intensity: weak - medium - pronounced

Fruit character: groups as for nose

Alcohol   : light - medium - high

Length   : short - medium - long

C O N C L U S I O N S

Quality   : faulty - poor - ordinary - good - outstanding

Maturity   : immature - youthful - mature - declining -
                                                      over-mature
Age/vintage   : unspecified wines

Provenence   : unspecified wines - grape variety, region

Commercial value : specified wines

The original Systematic Approach to tasting, as suggested by the author in 1987

The totality of tastes and flavours in a wine is complex, so it is essential that tasting be approached in a systematic way. The best way of doing this is to use the Systematic Tasting method as originated by the author in conjunction with Maggie McNie MW in 1987. The intention is that the left-hand column of tasting parameters should be memorised and should be used for the construction of every tasting note. This ensures that nothing is missed, and the note takes on a standard format. The descriptors in the right-hand side of the chart are suggestions only, and can be used at discretion. As tasting abilities develop, different descriptions will be used, but it should always be borne in mind that what is written should be intelligible, both to the author and to anyone else reading the note. Some of the best notes have been some of the shortest.

The Wine & Spirit Education Trust (WSET) has developed this approach to tasting and made it compulsory to use the words on their Systematic Approach to Tasting (SAT). Writing notes for an examination is a technique that has to be learned and adhered to, and this applies even at Master of Wine level. Tasting exams are marked by human beings and unless there is a rigid discipline it becomes impossible to mark fairly. Once the exam has been passed, poetry can develop!

Tasting the wine

Following the systematic approach step-by-step, it becomes much simpler to examine each aspect of the structure of the wine in turn, in the same order every time.

The left-hand column should be committed to memory and should be followed every time a wine is tasted. In this way tasting becomes a routine, with nothing forgotten.

The descriptive words suggested in the right-hand column are useful when developing a tasting technique, but should not be regarded as essential. Any descriptor can be used, provided it is meaningful. Attempts should not be made to copy entertainers on TV with amusing similes such as, "Smells just like a pair of old boots drying by the side of the stove on a wet November evening!"

Appearance

clarity – intensity – colour – other

Always hold the tasting glass by the stem and not the bowl, so that the glass remains clean. The colour of white wines is judged by the core colour alone; it is futile to describe the rim colour as 'watery'. All white wines have a watery rim!

With red wines the comparison of rim colour to core colour is important because it is on the rim that the signs of maturity are best judged. In fact, many red wines have such a deep colour that these can be described only as 'black' or 'opaque'. With these wines the observation of the rim colour is the only indication of the true nature of the appearance, and it is the main indicator of the degree of maturity of the wine.

Many people make a great deal of fuss about the 'legs', the tear-like drops that fall down the glass after swirling. These are of little importance because many wines show this phenomenon that is caused by the alcohol and glycerol that are present in all wines.

Nose

condition – intensity – development – character

Swirl the glass gently to release the aromas and to oxygenate the wine. When nosing (smelling) a wine, do not take a vast sniff but just let the aroma molecules drift up the nose so that they come into close contact with the olfactory organ which is situated at the top of the nasal cavity.

The moment of bringing the glass to the nose is of critical importance because it is what could be called a 'Damascus experience', when all is revealed or not, as the case may be! The time spent looking at the wine does not tell us much, but the first bouquet of aromas brings a huge packet of information which has to be recognised and analysed in a fleeting moment.

So important is this moment that all other senses should be shut off: eyes should be shut, ears closed and the body in total relaxation. The reaction to this moment can be excitement and pleasure or total devastation, according to the way in which the brain has interpreted the signals. If it is the latter, the only advice is to practice again and again, to build up a library of aromas in the brain that will give positive signals relating to the origins of the wine.

Palate

sweetness – acidity – tannin – body – intensity of flavour – character – alcohol – length

Take a moderate amount of wine into the mouth and swirl it around. Then, with the head leaning forwards, suck a stream of air through the wine to release more of the aromas that can reach the olfactory bulb via the retro-nasal route.

This movement is important because the taste receptors are situated in various parts of the mouth and not just on the tongue. The main sweet receptors are concentrated near the tip of tongue: hence this is the first sensation to be noticed. Acid receptors are nearer the sides of the tongue which results in a tingling sensation after tasting a wine with high acidity.

Bitterness is detected in various parts of the mouth and should be differentiated from the dryness due to polyphenols, which is caused by a reaction with the proteins in the saliva. The proteins are being precipitated in exactly the same way as in a fining operation, with the result that the mouth loses its lubrication.

One of the more advanced decisions that has to be made is whether the tannins are ripe and mature as in warm climate wines, or green and harsh as in wines grown in cool climates. The combinations can be quite complicated: a high level of ripe tannins, or a low level of green tannins. Ripe tannins are smoother and rounder, whereas green tannins have a greater drying effect on the mouth.

Finally, after spitting or swallowing, good wines have a lovely, lingering finish in the mouth, where the tastes and flavours should gradually die away harmoniously. Wines of lower quality often change their balance, leaving one or other of the characteristics out of balance. The length of the finish is often a good indicator of the quality of the wine.

Conclusion

quality – maturity – vintage – origins – price

An ability to judge the intrinsic quality of a wine is the most important attribute of a good taster. This is true even of the person who has just bought a bottle on promotion at the local supermarket. Was it a good buy? Was it enjoyable? Would you go back for another bottle?

Drinking - A few personal tips:

*   Drink in moderation regularly. Half a bottle of wine per day has been shown to be positively good, especially if it is red.

*   Don't binge-drink. This has a devastating effect on the body and should be avoided at all times.

*   Vary the style of wine you drink. Search for well-made wines at low price for regular drinking. Keep expensive wines for special occasions, even if it is just the weekend. You will then appreciate the difference.

*   Learn how to say 'Cheers!' in each country before you visit.

Old English	Wassail
Gaelic	Slainte
Scottish	Slainte Mhor
Welsh	Iechyd da
French	A votre santé
German	Prost
Italian	Salute or Cin-cin
Spanish	Salud
Flemish	Gezondheid
Hungarian	Egészségedre
Danish, Norwegian, Swedish	Skål
Finnish	Kippis
Bulgarian	Na zdrave
Latin	Sanitas bona
Zulu	Oogy wawa

Chapter 22

QUALITY ASSURANCE

There is nothing so useless as doing efficiently that which should not be done at all.

Peter Drucker, 1909-2005

Quality succeeds: the lack of quality will fail. All the art and science of wine making is wasted if there is no system for assuring the preservation of the potential that is being realised during the production of the wine.

It is important to differentiate between quality control (QC) and quality assurance (QA).

Quality control is a 'hands on' process of monitoring and controlling all parameters that affect the quality of the product at all stages of production, from the planting of the vineyard to the storing of the bottled wine. QC is not just something that technicians perform in a laboratory, but should be a sequence of observations that are applied by everyone who has any part to play throughout the process. Production personnel are responsible for quality, not QC.

Quality assurance is the totality of all the management actions and procedures that set out to achieve this high standard, and this will incorporate quality control. QA is a management concept, covering the manner in which a business is organised, so that the quality of its product is assured at all stages. As applied to a wine business, good QA will ensure that the original potential of the wine is not lost on the way to the bottle.

There are several models on which to base a quality assurance programme, models that help by providing a standard base from which to work. In the food industry the most important of these is Hazard Analysis and Critical Control Points (HACCP), not least because it is required by law. This is the best base from which to start. Having developed this, the next step might well be to aim for registration to ISO 9001, a quality management system which has proved to be within reach of any well-organised company, however small. Even better for any company in the food and beverage industry would be ISO 22000, the food safety standard.

Following on from these, there are many further steps to improvements in quality management, such as Total Quality Management (TQM), which introduces the concept of a business being a succession of processes, where everyone is both supplier and customer in a chain of events. In this case, customer satisfaction is required throughout the business, and not just from the ultimate external customer. Or there is the Japanese concept of Kaizen, where continuous improvement in small steps is demanded. The ultimate standard in Europe is the European Quality Award given annually to companies that satisfy a very stringent set of requirements.

There are many ways in which the attitude towards quality can be raised. The important factor is that it should be an integral part of any business plan, being equally important to both profit and expansion.

Hazard analysis and critical control points (HACCP)

Hazard Analysis and Critical Control Points (HACCP) is a system of food control developed in the 1960s jointly by the Pillsbury Company, the US Army Laboratories and NASA, to ensure the safety of foods for the American space programme. It is essentially an exercise in preventative action and, although originally aimed at preventing microbiological contamination, it can be used for any aspect of manufacture and can deal with any hazard, whether physical, chemical or microbiological. It can be made as broad or as detailed as desired, even to incorporating factors affecting quality. The danger is in becoming too detailed, when the process becomes very long-winded and tedious. The important criterion is getting the guidelines correct in the first instance, which requires experience.

In Europe, the Food Hygiene Directive 93/43/EEC of 14 June 1993 required all member states to change their legislation relating to food safety. The details were published in Regulation (EC) No 852/2004 on the hygiene of foodstuffs.

The definition of HACCP in the Codex Alimentarius is "A system which identifies, evaluates and controls hazards which are significant for food safety", and a Critical Control Point is "A step at which control can be used to prevent or eliminate a food safety hazard or reduce it to an acceptable level." Making sure that food is safe to eat is a good starting point, but many companies have found this principle so useful that they widen the scope to incorporate other hazards that might affect the quality and the legality of the product. These can be identified within the system by using additional letters such as SCCP for safety, QCCP for quality and LCCP for legality. It is worth looking at this in greater detail.

Principles of the HACCP system

According to the Codex Alimentarius there are seven steps or principles in the system:

1. Conduct a hazard analysis.
2. Determine the Critical Control Points (CCPs).
3. Establish critical limits.
4. Establish a system to monitor the control of the CCP.
5. Establish the necessary corrective action.
6. Establish procedures to verify that the system is working.
7. Establish documentation and recording.

The process

1.   Assemble the HACCP team.   This should consist of representatives from all parts of the production process, and must include somebody who has had training in the application of HACCP. For a winery this might include a viticulturalist, an oenologist, a winery manager, an experienced quality control technologist, a bottling supervisor, a dry goods buyer and an engineer. It is important that there should be an input from all aspects of the operation.

2.   Set the scope of the investigation and describe the product. It may well be too difficult to take the business as a whole, so it should be broken down into manageable parts, which for a winery might be managements of the vineyard, making the wine and the bottling of the wine.

3.   Identify the intended use of the product, which might be for sale in bulk, for bottling for export, or for home sales.

4.   Construct a flow diagram of the process, from the gathering of the grapes to the wine in the bottle. The purpose of this is to ensure that every step in the process has been considered and to help to concentrate the mind on each step in turn.

5.   Confirm this by a site visit, if necessary.

6.   List all the hazards associated with each step, and there could be many. This is the major part of the exercise, and is the point at which decisions have to be made whether to incorporate quality and legality hazards, or whether to keep it to its original purpose of identifying food safety hazards.

7.   The next stage is probably the most difficult to define, as this is when the Critical Control Points have to be identified. One method is to assess each hazard according to the probability of it happening and the gravity of the effect if it should happen. The decision is then based on a table as below.

Gravity / Probability	Mild	Medium	Serious
Frequent	YES	YES	YES
Sometimes	NO	YES	YES
Rare	NO	NO	YES

Another method is by the use of a 'Decision Tree' as set out in the Codex Alimentarius. This is essentially asking the following five questions for each hazard:

Q1 Do preventative control measures exist?
Q2 Are these control measures necessary at this step?
Q3 Will these control measures eliminate or reduce the hazard to an acceptable level?
Q4 Could contamination occur to unacceptable levels?
Q5 Will a subsequent step eliminate this danger?

Possibly a combination of the two will be used and the conclusion reached will depend to a large extent on the expertise and professional knowledge of the members of the HACCP team.

8.   The critical limits for each CCP are set. These could include measurements of temperature, pH, or analytical limits.

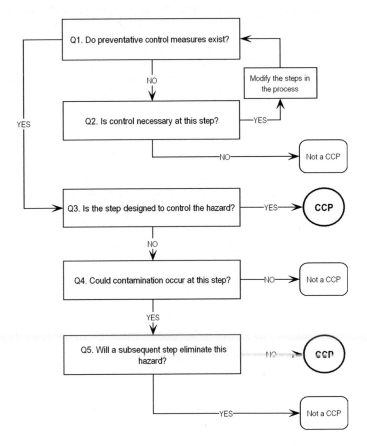

The Decision Tree, as defined in Codex Alimentarius

9.   A monitoring system is set up so that the CCP can be observed.

10. Corrective actions are established to keep the CCP within the critical limits.

11. A system of verification is set up to confirm that the CCP is under control.

12. Documentation and records are set up.

All of this is really nothing more than the formalisation of a process that has been applied for many years by any operator with a grain of common sense.

The application to a winery

There are huge differences in the application of the HACCP system in wineries, with the number of CCPs ranging from as few as three to more than twenty, when around ten would be expected.

Much management time has been used in the process of developing this system in each winery, when it should be possible to create a procedure that can easily be adapted to individual requirements. The process is much the same in all wineries: grapes are picked, delivered and processed, and the juice is fermented, clarified and bottled. Wine, fortunately, is comparatively easy to handle, and is safe in respect of human health, as no harmful organisms are able to survive, due to the low pH and the presence of alcohol. The principal problems relating to human health have been the use of illegal sprays in the vineyard (or excessive amounts of legal chemicals), metal contamination in the winery, too much sulphur dioxide and the incorporation of foreign bodies during bottling. Therefore it is not surprising that the number of CCPs is comparatively low.

It is interesting to note that the New Zealand Food Safety Authority states in its HACCP application document that "No CCP was identified for the production of grape wine." However, this is on the understanding that all winegrowers abide by the Code of Practice laid down in the New Zealand Winegrowers' Wine Standards Management Plan.

The following scheme is offered as a starting point for the development of a HACCP plan for any winery that has not yet embarked upon such an exercise. On the following pages are shown the process flow chart in two sections: wine making, followed by treatments and bottling. The hazards associated with each step are then listed, and each is analysed according to the 'decision tree' to decide whether it is critical to the process. The result of this analysis is the production of a table of Critical Control Points, as shown on page 277.

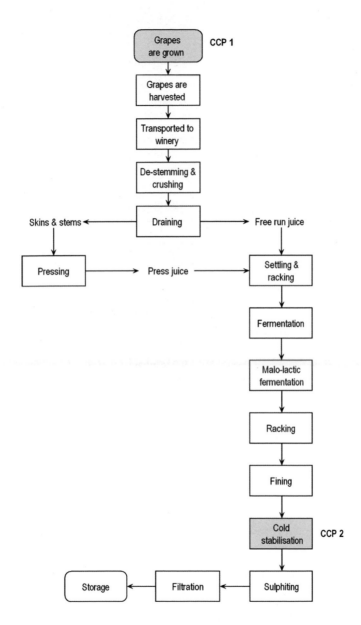

Process flow chart for the production of wine

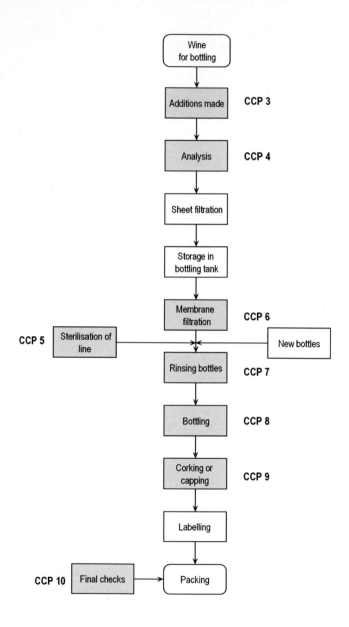

Process flow chart for the finishing and bottling of wine

TABLE OF CCPs

Stage	CCP	Hazard	Preventive measure	Critical limits	Surveillance	Corrective measure	Records
Vineyard management	1	Illegal sprays Pesticide residues	Vineyard manual	Nil present	Treatment plan Log books	Training	Log books
Cold stabilisation	2	Contamination by coolant	Regular maintenance	Nil present	Maintenance reports	Analysis if leak suspected	Log books
Sulphiting	3	Exceeding limit	Analysis	As laid down in regulations	Checking of records	Training	Analysis records
Analysis	4	Incorrect results	Ring tests Standard wines	Experimental accuracy	Checking of records	Training	Analysis records
Sterilisation of bottling line	5	Machinery not sterilised	Microbiological testing	Nil organisms	Scrutiny of results	Revise sterilisation procedure	Production records
Filtration	6	Fermentation in bottle	Integrity testing	Nil organisms in bottle	Checking of result	Training	Print-out
Bottle rinsing	7	Foreign bodies not eliminated	Regular checking of water jets	Nil foreign bodies	Spot-check testing	Improve maintenance	Bottling records
Filling of bottles	8	Pieces of broken glass	Close attention to filling machine	Nil foreign bodies	Visual checking	Maintenance & training	Bottling records
Corking or screw-capping	9	Foreign bodies: glass or cork	Close attention to machinery	Nil foreign bodies	Visual checking	Maintenance & training	Bottling records
Final checks	10	Nonconformance	Careful scrutiny	Conforms to specification	Checks by manager	Wine quarantined	QA records

Interpretation

The source of the difference found from winery to winery lies in the interpretation of the HACCP principle. Common sense has to be applied to any system of quality management in order to obtain a sensible result. All such systems are aimed at improving a business and making it both easier to run and more efficient, as well as guaranteeing a good product. Wine is a foodstuff and therefore must also be safe to consume.

It is not possible, or indeed sensible, to try to monitor some of the CCPs in the wine process. How can the absence of illegal pesticides be monitored? Every pesticide analysis is very expensive and difficult to perform. How can the absence of particulate matter in the bottle be confirmed? It would be possible to check the occasional bottle by tipping the contents through a filter and checking for particles, but what would this tell us?

The whole purpose of HACCP is to focus attention on the potential dangers and to institute preventative action. The way this is done is to apply common sense, and to review the processes accordingly, without putting the company out of business.

ISO 9001:2008

ISO 9000 is one of many quality management systems that assists a business to manage the quality of its product or service in a systematic manner. It is based on the total documentation of the business procedures, a tight internal control of the system and an on-going assessment by an accredited body.

It began life after the Second World War (1939-1945) in munitions factories, where armaments were exploding prematurely, with somewhat dire results. It was realised that formal procedures were required to ensure consistency of production, which led to the development by NATO of the Allied Quality Assurance Publications (AQAP) series of standards. These were developed into standards for the wider field of engineering, and in 1979 were published by the British Standards Institution as BS 5750. The success of this British standard became world-wide and it was adopted by the International Standards Organisation as ISO 9000.

In the year 2000, a new revision was issued, which removed a lot of the criticism of the previous versions issued in 1987 and 1994. For the first time, the system is based on processes rather than procedures. Many of the mandatory documented procedures have been dropped, with only four remaining. At last, it has been realised that a person can be trained to do a job, rather than relying on a written procedure.

The other main changes are an emphasis on continuous improvement and a major focus on customers, who are, after all, the most important element of any business.

Some would say that ISO 9000 is bureaucratic and causes a vast increase in paperwork, general workload and costs. This could and does happen, but it is entirely dependent upon the manner in which the system is installed: if done properly, the result is the exact opposite.

Preparations for installation of the systems required by ISO 9000 give an excellent opportunity to examine, review and improve existing systems. In the course of this, people are involved and motivated, which automatically improves communication.

The result, when all the processes have been identified, is that a new-found clarity emerges, where everyone knows who does what, how they do it and whose responsibility it is.

Both internally and externally, a commitment to quality is demonstrated, which raises morale and impresses customers. Costs are reduced, both by increased efficiency and by reduced wastage, because the structures that are put in place are aimed at getting it right first time. ISO 9000 was designed to help a business: if it is proving a hindrance, it has been wrongly installed.

The twenty requirements of the old ISO 9000 have been grouped under five main headings:

1. Quality management system
2. Management responsibility
3. People
4. Product realisation
5. Measurement, analysis and improvement

The standard is wider ranging than the old versions, but is easier to manage, and is moving steadily towards real business management.

Before granting registration, the certification body, which should itself be accredited by the national accreditation body (which in the UK is UKAS - the United Kingdom Accreditation Service), carries out a detailed assessment of the entire system. Return visits are made periodically for a long as the registration remains valid.

In summary, ISO 9000 is common sense written down on paper. It is a system that enforces a discipline. Handled correctly, it does not stifle development or prevent change. It is meant to be an aid to the achievement of quality: if it is not, then it has been badly installed and should be re-visited.

ISO 14000:2004

Quality assurance nowadays is not directed simply at quality: we have to be conscious of the effect we have on the environment. ISO 14001:2004 is a management tool enabling an organization of any size or type to:

- identify and control the environmental impact of its activities, products or services;

- improve its environmental performance continually;

- implement a systematic approach to setting environmental objectives and targets, to achieving these and to demonstrating that they have been achieved.

Fulfilling these requirements demands objective evidence which can be audited to demonstrate that the environmental management system is operating effectively in conformity to the standard.

ISO 22000:2005

This is the latest of the ISO series of standards appropriate to the wine industry and takes the approach of ISO 9000 as a management system, incorporates the hygiene measures of Prerequisite Programmes (PRPs, otherwise known as Good Manufacturing

Practice), and adds HACCP principles and criteria. This gives the best of the management system approach added to the well established PRP and HACCP programmes.

In theory, it should render obsolete all supplier audits and the BRC Global Food Standard as described below. However, this seems unlikely, as these are well established and widely used.

Supplier audits

The Food Safety Act 1990, in particular, has spawned a whole new business of supplier audits. Specialised companies have grown up, employing rafts of auditors, some better than others, some fully qualified, some very inexperienced. The United Kingdom Accreditation Service (UKAS) is flourishing.

This has been brought about by the concept of 'due diligence', whereby the supplier of any foodstuff has to demonstrate that they "took all reasonable precautions and exercised all due diligence" to ensure that the foodstuff they are supplying is safe. One of the steps that can be taken is to visit all suppliers of any component or finished product and carry out an audit.

Three problems have arisen from this practice:

- Many audits have been based entirely on compliance with set procedures designed for product safety, and have had no bearing on product quality.

- Suppliers have had to endure multiple audits, because each retailer insists on carrying out their own audits.

- Audits have been carried out by people who, although fully qualified as auditors, have no specialist knowledge of the product being audited. This results in unnecessary and irritating standards being imposed.

The converse is that if the audit is conducted by somebody with wide knowledge of the industry sector, then positive good can result and best practice can be promulgated.

The BRC Global Food Standard

The British Retail Consortium (BRC) is the lead trade association representing a wide range of retailers, from the large multiples and department stores through to independents, selling a wide selection of products through centre of town, out of town, rural and virtual stores.

In 1998 it published the first edition of what has become its Global Food Standard which sets the benchmark for food safety management systems, laying down criteria against which companies can be assessed and so allowing purchasers to buy with confidence. It is used for auditing any supplier of branded food products to any of its members.

It was designed to cover all foodstuffs, from the highest risk, such as dairy, fish and meat products, to the lowest, such as wines and spirits. It incorporates all of the requirements of HACCP and elements of quality management and factory environment standards, and suppliers are subject to an audit by an accredited auditor.

The original intention was to eliminate the multiple audits that suppliers were having to endure. Unfortunately, this ideal has not been achieved, as many of the major retailers still have the urge to see for themselves that everything is in order.

The all-encompassing requirements have also caused a problem in some wineries, as inexperienced auditors have insisted on conditions quite unnecessary for the making and bottling of wine. The important criterion is that the audits should be carried out only by those who are expert in the field in which they are auditing, and this has not always been the case. There are many wine producers who rage against the standard, having been forced to impose all manner of ridiculous conditions in their wineries in order to satisfy the apparent requirements.

On close examination, it would appear that this is not the fault of the standard itself, but of certain auditors who want to apply the strict regime of a high-risk poultry or dairy production plant to the very safe production of wine. There are many cases where considerable sums of money have been involved in instituting totally unnecessary changes to plant and personnel. In these circumstances, communication should be made with the certification company or to the BRC itself.

Nevertheless, this standard has been very successful. It is used by certification bodies in 23 countries across 4 continents, and is now in its fifth edition.

Business Excellence Model

The ultimate system for Europe is the Business Excellence Model, as promoted by the European Foundation for Quality Management (EFQM) that was set up in 1988 by fourteen chief executives of leading European companies. It has since grown to over 750 members across Europe and has used the Business Excellence Model as a means of enhancing the effectiveness and efficiency of participating companies. Similar schemes have been launched in other parts of the world, notably the Baldridge Awards in the United States.

The model is an all-encompassing system that covers nine aspects of business management:

1. Leadership
2. People management
3. Strategy & planning
4. Resource management
5. Quality systems & processes
6. People satisfaction
7. Customer satisfaction
8. Impact on society
9. Business results

Unlike other quality systems, it is not based on compliance, but is rather a way of life centred around continuous improvement and self-assessment. When the management team feels the time is right, the company can put itself forward for one of the European Quality Awards, which are given each year to the organisation judged to be the best in its category.

For further information:

For HACCP:

>Codex Alimentarius Food Hygiene Basic Texts.
>*Food and Agriculture Organisation of the United Nations/World Health Organisation,* ISBN 92-5-104021-4

For ISO standards:

>*International Organisation for Standardisation (ISO)*
>www.iso.org/iso/home.htm

For BRC Standard:

>*The British Retail Consortium (BRC)*
>www.brcglobalstandards.com

For Business Excellence:

>*The European Foundation for Quality Management (EFQM)*
>www.efqm.org/en/

Chapter 23

LEGISLATION & REGULATIONS

La liberte est le droit de faire tout ce que les lois permettent
(Freedom is the right to do anything the laws permit)

Montesquieu, 1689-1755

The principles of European law operate at two different levels:

- Directives demand that each country in the European Union should adapt its own laws to follow the rules laid down in the Directive.

- Regulations are legislation that is mandatory for all countries for immediate application.

Despite statements to the contrary, European wine legislation gets more and more tortuous and complicated as time goes on. New pieces of legislation appear on top of existing laws, some previous regulations are repealed, some are expanded. It is extremely difficult to obtain the entire raft of legislation pertaining to a particular product, as the various items are widely spread over numerous regulations and directives. It would be wonderful to be able to start from scratch with a blank pad of paper and to put all the laws and regulations into a sensible and logical order, but that will not happen!

This chapter attempts to list the most important parts of European wine legislation, with a brief resume of the major purpose of that law. Also included are some of the UK regulations that apply specifically to wine.

The one aspect of European law that has improved is access to the written documents. They are all obtainable free of charge from

http://eur-lex.europa.eu/RECH_menu.do .

Click on CELEX number and follow the instructions. Good luck!

In the UK the European legislation is enforced through the issue of Statutory Instruments entitled The Common Agricultural Policy (Wine) Regulations under various SI numbers. These change frequently, so it is necessary to search for the latest version.

Regulation 479/2008
The Common Organisation of the Market in Wine

This is the so-called 'framework' regulation that sets out the general conditions for the wine market in Europe. The previous regulation (R1493/1999) has been repealed as it was no longer appropriate.

Its titles cover:
- Support measures
- Regulatory measures, including labelling rules
- Trade with third countries
- Production potential
- General provisions

It is not an easy document to understand because it contains a strange mixture of fiscal, oenological and labelling requirements.

The main changes are that the terms 'Table wine' and 'Quality wine produced in a specific region' disappear and are replaced by PDO or PGI as detailed in Regulation 607/2009 (see below). It is now also permissible to include vintage and grape variety on those wines that would have been labelled Table Wine in the old order.

Community symbols for PDO and PGI wines

The basic rules for labelling remain the same, with compulsory and optional information:

- The category of the grapevine product
- PDO or PGI, if applicable (or using the traditional terms, AC etc.)
- Actual alcoholic strength, expressed as % vol
- Indication of provenance

- Name and head office address of the responsible packer in the EU
- For third country wines, the name and address of the importer

This is not the totality of compulsory information as other elements are enshrined in other parts of the legislation, e.g. Nominal volume and "Contains sulphites".

The optional information includes:
- Vintage year
- Grape variety
- Indication of sweetness
- The Community symbol indicating PDO or PGI
- Production method
- Geographical unit

Regulation 606/2009 Detailed rules

This regulation supplements R479/2008 by supplying the details of the winemaking practices. As above, it is a very strange mixture of oenological practices and a few methods of analysis, mixed up in annexes and appendixes. Annex 1 is useful as a list of all permitted oenological practices.

Regulation 607/2009 More detailed rules

This regulation introduces the concept of Protected Designation of Origin (PDO) and Protected Geographical Indication (PGI). These terms can be used instead of the traditional terms such as AC or DOC, but in fact the old terminology will still be used, and these new terms will merely confuse the issue even more.

Some of the detailed labelling rules are incorporated in this regulation.

Regulation 1991/2004 Declaration of allergens

This regulation amends R753/2002 (driven by D2003/89) by making compulsory for wines containing more than 10 mg/litre of sulphur dioxide (which means in effect all wines) the declaration "Contains sulphites" or "Contains sulphur dioxide", or even "Contains sulfites". This applies to all wines bottled after 25 November 2005. Other allergenic ingredients may be added to the list depending on outcome of Scientific Opinions of European Food Safety Authority (see p.152).

Directive 1989/396 Lot marking

Under this directive all foodstuffs must carry a lot mark to enable a product re-call to be carried out quickly and efficiently, if necessary. This must commence with a capital L and must be a number unique to a batch. The size of a batch is not defined; it must merely be identifiable. It might be a day's bottling, but equally it might be the entire vintage. It is sensible, however, to keep the batch as small as possible in case of any re-calls.

Most bottlers of wine use a format of four digits, where the first digit is the last digit of the year and the other three are the day of the year according to the Julian calendar. For example, 31 January 2010 would be L0031.

Regulation 178/2002
Principles and requirements of food law - Traceability

This wide-ranging regulation contains a principle that has had a major effect on the way that all sections of the wine trade operate. Article 18 states that "The traceability of food . . . shall be established at all stages of production, processing and distribution." Wine is food, and is not exempt. This is enforced in the UK through General Food Regulations SI 2004/3279.

The effect of this is that all wineries will have to install systems that enable traceability of every batch of bottled wine back through all the processes and blends to the original grapes and even to the vineyard site from whence they were harvested. This is a major problem for the producers of large volumes of table wine, where the final blend might well consist of numerous components from all parts of the country.

Directive 2000/13 Labelling, presentation & advertising

The labelling, presentation and advertising of wine is covered by the wine sector regulations listed at the beginning of the present chapter. The only extra parameter that falls within the above directive is ingredient listing, which has received many derogations, on the grounds that the addition of ingredients is tightly controlled by the wine regulations. Very few ingredients are used, mostly leaving no residue in the wine because they are processing aids and are removed during fining.

Weights & Measures (Packaged Goods) Regulations 2006

All pre-packaged foodstuffs are now packed according to the European average system of quantity, which is different from the old British system of minimum quantity. Under the British system, every bottle of wine had to contain at least the quantity stated on the label (the nominal quantity). With the average system, the entire batch of wine is regarded as the unit, i.e. the average of all the bottles in the batch must be not less than the nominal quantity, but each individual bottle can be above or below the nominal, within certain limits.

The lower limit is governed by the tolerable negative error (TNE) which for a 750 ml bottle is 15ml. Up to 2½ % of the batch can have negative errors greater than this, but no bottle is allowed an error greater than twice the TNE, which is 30ml. Therefore, a 750 ml bottle of wine is allowed to fall as low as 720ml, but because the average for the batch must be at least 750ml, some lucky people will get more. In practice, wine filling is much more even than this and most bottles will hold between 740ml and 760ml.

When this average system is certified as having been used by a packer, and it has been checked by the local authority for weights and measures, the e-mark may be put on the label. If the wine has been bottled within the EU, the shipper of such wine need not be concerned with any further checking. The responsibility for correct filling lies with the authority in whose area the wine was packed. For wines from third countries, the importer should satisfy himself that proper testing has taken place, with suitable documentary evidence.

Weights & Measures
(Specified Quantities)(Pre-packed Products) Regulations 2009

This regulation will give greater freedom for the prescribed sizes in which wine can be sold in that control will apply only to those sizes within the range 100 ml to 1500 ml. This will open the way for small quantities of wine to be sold as tasting samples, e.g. 25 ml, thus correcting an anomalous situation that has prevented the sampling of wines before the purchase of a standard bottle.

However, at the time of printing (July 2010) this regulation has not been enacted.

Within the controlled range the prescribed sizes will be:

Wine: 100, 187, 250, 375, 500, 1000, 1500 ml
Sparkling wine: 125, 200, 375, 750, 1500 ml
Liqueur wine: 100, 200, 375, 500, 750, 1000, 1500 ml
Vin jaune: 620 ml

Food Safety Act 1990

The Food Safety Act has become synonymous with the phrase 'due diligence', which in some ways is unfortunate. Firstly, this phrase is but one small part of an Act which introduced wide-ranging changes to food law, and secondly because it is associated with a defence against prosecution rather than positive action. The 'due diligence' phrase appears in chapter 16 section 21(1) where it states that it shall "be a defence for the person charged to prove that he took all reasonable precautions and exercised all due diligence to avoid the commission of the offence".

Food Safety (General Food Hygiene) Regulations 1995

One of the most interesting concepts introduced is Hazard Analysis of Critical Control Points (HACCP), which is dealt with in chapter 18. This procedure has been incorporated in European hygiene legislation and should, therefore, be addressed in all EU member states.

Glossary

Acid   A substance with a sour taste and a pH of less than 7. Contains excess hydrogen ions and reacts with a base to form a salt.

Absorption   Molecules of a substance enter the body of a solid (*cf.* adsorption).

Adsorption   The adhesion of molecules to the surface of a solid.

Aerobic   Requiring the presence of oxygen. In aerobic exercises, the muscles are never deprived of oxygen.

Albumin   A water-soluble protein found in egg whites etc. Coagulates with heat. Used for fining.

Aldehyde   A substance with the –CHO group, typified by acetaldehyde, a strong-smelling compound produced during the oxidation of wine.

Alkali   A substance with a pH greater than 7 and typified by the –OH group. Forms a salt with an acid. (*Syn.* Base)

Allergen   A substance that can cause an allergy.

Amino acids   An organic acid containing the $-NH_2$ group. The building blocks of proteins. Synthesised by living cells or obtained from the diet.

Anaerobic   The absence of oxygen.

Anthocyanins   Soluble polyphenolic pigments in plants, ranging from red to blue. Colours change with pH.

Antioxidant   A substance that minimises the effect of oxidation.

Aqueous   Relating to water.

Archimedes screw   A broad-threaded screw encased in a tube. Originally used to raise water.

Atmosphere   The mass of air surrounding the earth, consisting approximately of 78% nitrogen and 21% oxygen. Also a unit of pressure of 14.7 pounds per square inch, or 760 mm of mercury.

Atom   The smallest unit of a chemical element that has the properties of that element.

Atomic weight The mass of one atom of an element.

Bacterium Single-cell micro-organism, smaller than a yeast. *Pl.* bacteria.

Bar Unit of pressure of 1 atmosphere, or 14.7 pounds per square inch (psi), or 760 mm of mercury.

Barrique   A barrel used widely around the world. Capacity 225 litres.

Bentonite An aluminosilicate clay $Al_4Si_8O_{20}(OH)_4.nH_2O$ that swells in water and has powerful properties of adsorption.

Biochemical Involving chemical reactions in living organisms.

Bronze An alloy of copper and tin, of varying proportions.

Buffering effect The ability of dissolved salts to affect the pH of a liquid.

Carcinogen A substance that can initiate cancer.

Casein A colloidal protein occurring in milk. Plays a part in the prevention of curdling. Used for fining wine.

Casse From the French: breakage. A solid breaking out of solution. A precipitate.

Catalyst A substance that enables a chemical reaction to proceed at a faster rate, but does not take part in the reaction.

Chapeau   The floating layer of grape skins produced during fermentation of red wines.

Chlorophyll The green coloured substance in plants that enables the process of photosynthesis.

Colloid   A dispersion of small particles in a liquid, the particles being too small to be visible to the naked eye, but large enough to be visible under the microscope.

Complex A substance formed by the loose association of two or more chemical compounds.

Compound A substance that can be decomposed into its constituent elements by means of a chemical reaction.

Decomposition Breaking up into constituent parts.

Density The mass of a substance per unit volume.

Distillate The liquid collected from the condenser during a distillation process.

Element A substance that cannot be decomposed into simpler substances by chemical means.

E-number The code number given by the EU to food additives.

Enrichment The addition of sugar to grape must to increase the alcoholic strength of a wine after fermentation.

Enzyme Biochemical catalyst, mostly proteins.

Epoxy resin A tough, resistant paint which sets by polymerisation of two components.

Esters Fragrant compounds formed by reaction between an alcohol and an acid, e.g. ethyl acetate.

Fermentation A biochemical reaction involving enzymes.

Flocculate A loose deposit formed from aggregated cells.

Fortification The addition of ethanol to increase the alcoholic strength of a wine.

Free-run juice The juice that runs out of a crushed grape without the application of pressure.

Fructose A sugar occurring in fruits, including grapes, along with the sugar glucose, its structural isomer. Molecular formula $C_6H_{12}O_6$.

Glucose Occurs also as the major sugar in the blood of higher animals. Also known as dextrose. Molecular formula $C_6H_{12}O_6$.

Heat exchanger A device for rapidly raising or lowering the temperature of a liquid.

Hydrolysis Chemical decomposition by water.

Ion An atom or molecule that has lost or gained electrons and thus possesses an electrostatic charge.

Intracellular Within the confines of a cell.

Isinglass A pure form of gelatine obtained from the swimbladder of the sturgeon.

Isomers Molecules that contain the same atoms, but in a different structural formation.

Ketone Compounds typified by acetone. In wines generally one of the products of oxidation.

Kieselguhr A diatomaceous earth used as a filter aid.

Lees The solid deposit at the bottom of a vat.

Leguminous plants Plants whose seeds are contained in pods. Most harbour nitrifying bacteria on their roots.

Lipoprotein A substance that contains both fat and protein.

Maceration Steeping solids in liquids to soften them and to aid the extraction of compounds.

Marginal climate A climate that barely exceeds the minimum requirements for growth. Can produce high quality.

Metabolism The processing of a substance by a living organism.

Meta-stable Having a small margin of stability.

Microbiology The science of microscopic life forms.

Microclimate The local climate of a small area.

Microorganism Organism of microscopic size: yeast, bacterium, virus.

Minerals Naturally occurring substances containing important salts of metals.

Molecular weight The mass of one molecule of a substance. Calculated from the sum of the individual atomic weights.

Molecule The smallest particle of a substance that retains the characteristic properties of the substance.

Must Unfermented or partially fermented grape juice, with or without the skins.

Nitrifying bacteria Bacteria that convert atmospheric nitrogen to nitrogen compounds, e.g. nitrates, thus 'fixing' nitrogen.

Noble rot *Botrytis cinerea, pourriture noble, Edelfäule.*

Oenologist A wine scientist (Gk *oinos* wine). Also sp. enologist.

Organic Relating to living organisms and based on the chemistry of carbon.

Oxidation Reactions involving the combination of molecules with oxygen.

Pathogen An organism that causes disease.

Pectin A gelatinous substance that binds together adjacent cell walls in plant tissue.

Pectinolytic enzyme A enzyme that destroys pectin by hydrolysis.

Photosynthesis The process by which light energy is used to convert carbon dioxide to carbohydrates.

Phytoalexin A natural antibiotic that plants produce when under stress.

Planting density The number of vines per unit area.

Polymerisation The process by which molecules join together.

Polyphenols A large group of compounds including the tannins and the anthocyanins. Many are powerful antioxidants.

Potassium bi-tartrate HOOC.CHOH.CHOH.COOK The substance of most tartrate crystals. Also known as potassium hydrogen tartrate, cream of tartar.

Potassium metabisulphite $K_2S_2O_5$ A white powder and a useful source of sulphur dioxide.

Precipitate A solid that has been thrown out of solution.

Press juice The juice that is extracted from grapes by pressure.

Proteins Complex substances built up from amino acids. An important component of living tissue.

PVPP Poly-vinyl poly-pyrrolidone, a manufactured polymer (plastic). It is a gentle fining agent that removes phenolic compounds from wine.

Reduction The opposite of oxidation: a chemical reaction involving the removal of oxygen.

Salt A chemical compound formed by reaction between an acid and a base. Common salt or sodium chloride $NaCl$ is one example of a salt.

Solubility The degree by which a solid will dissolve in a liquid.

Stomata The openings in the surface of a leaf by which the plant breathes.

Sucrose $C_{12}H_{24}O_{12}$ The sugar of sugar beet and sugar cane.

Sugars A group of water soluble compounds of various degrees of sweetness.

Tannins Colourless polyphenols that impart a bitter taste to wine, tea and other foodstuffs.

Tartaric acid HOOC.CHOH.CHOH.COOH The most abundant acid in grape juice. The strongest of the acids of wine.

Titration The process of the determination of the concentration of a dissolved substance by the addition of measured quantities of a suitable reactant.

Viscosity The resistance to flow in a liquid.

Volatile Easily vaporised.

Yeast A single cell micro-organism that reproduces by budding. *Saccharomyces cerevisiae* is the variety used for the majority of wine fermentations.

Bibliography

Beyond alcohol: Beverage consumption and cardiovascular mortality. *Clinica Chimica Acta,* 1995, **237**, 155-187

Campbell, Christie, *Phylloxera*, Harper Collins, 2004

Clarke R. J. & Bakker J., *Wine Flavour Chemistry*, Blackwell, 2004

Ford, Gene, *The Science of Healthy Drinking,* Wine Appreciation Guild, 2003

Goode, Jamie., *Wine Science,* Mitchell Beazley, 2005

Hornsey, Ian, *The Chemistry and Biology of Winemaking,* The Royal Society of Chemistry, 2007

Jackson, R. S., *Wine Science,* Academic Press, 2000

Johnson, H & Robinson, J, *The World Atlas of Wine*, 5th edition, Mitchell Beazley, 2001

Rankine, Bryce., *Making Good Wine*, Pan Macmillan, 1989

Ribereau-Gayon et al. *Handbook of Enology,* Wiley, 2000

Robinson, Jancis (Editor), *Oxford Companion to Wine,* 3rd edition, Oxford University Press, 2006

Sandler & Pinder, *Wine: A Scientific Exploration,* Taylor & Francis, 2003

Skelton, S *Viticulture,* 18 Lettice Street, London, SW6 4EH, 2009

Micro-oxygenation – A Review
The Australian & New Zealand Grapegrower & Winemaker 2000

Wine Business Monthly - various papers
www.winebusiness.com

Wine, alcohol, platelets, and the French paradox for coronary heart disease. *The Lancet,* 1992, **339**, 1523-1526

Wine bottle closures: physical characteristics and effect on composition and sensory properties of a Semillon wine. *Australian Journal of Grape and Wine Research Vol.7, No 2, 2001*

Conclusion

The great fascination of the study of the technology of winemaking is the realisation that there is no single right way of doing it. At every stage there are options open to the winemaker. Decisions have to be made, very often with no scientific evidence to help. Good wine is made by the intelligent winemaker who knows the grapes, who knows the effects of the various treatments and who knows what is wanted of the ultimate wine.

The potential quality of wine is present in the grapes when they are picked. "Good wine is made in the vineyard." The application of the principles of good quality control and quality assurance will ensure that the potential of the grape is upheld until the wine is consumed.

Despite the help that science can give to the art of winemaking, there can be no doubt that the winemaker is an artist, moulding the transformation of grape juice into one of those rare things that can give both pleasure and health, and whose study gives infinite rewards.

Drink wine, and you will sleep well.
Sleep, and you will not sin.
Avoid sin, and you will be saved.
Ergo, drink wine and be saved.

Medieval German

Index

A

absolute filter 189
acacia 126, 147, **177**
acetaldehyde 59, **142**, 171, 242
acetals 73
acetic acid 4, 72, 138, 216, **238**, 255
acetic bacteria 88, 239
acetobacter 139, 169, 179, 215
acid rain 166
Acidex 65
acidification 63
acids 137
acid-sugar balance 57
activated charcoal 70, **152**
additives 165
aerobic 74
aerobic winemaking 32
agar 77
agglomerate cork 204
agraffe 114
albumin 150
alcohol 3, 6, **135**, 240
alcohol dehydrogenase 6
alcoholic strength 232
aldehydes 27, **142**, 171, 226
allergens **152**, 287
allergies 168
aluminium 198, 210
aluminium cans 198
aluminium foil 201
Alzheimer's 8
amino acids 28, 31, 60, 69
ammonium compounds 69
ammonium sulphate 70
amontillado 123
ampelography 9
añada 121
anaerobic maturation 226
anaerobic winemaking 32
anaerobiosis 99
anaesthetic 135
analysis 227

anthocyanins **26**, 59, 86, 94, 96, 101, 119, 125, 134
anticoagulant 6
anti-mutagenic 7
anti-oxidasic 170
antioxidants 6, 33, 60, 249
antiseptic 169
archimedes screw 48, 55
argon 34, **36**
aroma trap 68
aromatics 80
artificial cooling 79
asbestos 187
Ascomycetes 72
ascorbic acid 33, 107, **174**, 242, 243
aseptic 202
aseptic bottling **215**, 247
Aspergillus niger 156
aspiration method 242
Asti 82, **117**, 221
atmosphere 31
atomic absorption 245
Auslcsc 110
autolysis 114
autovinifier **94**, 119
average system 289

B

bacteria 4, 59, 72, 84, 86, 176, 198
bad egg 176, 198
bag-in-box **198**, 252
barrel fermentation 131
barriques 125, 128
basket press 49
bâtonnage 106, 131
Baumé 231
Beaucastel 97
Beerenauslese 110
beet sugar 65
bentonite 70, **151**
Bentotest 241
best before 252
β-glucans 179
bctaglucanasc 179
bidule 115

binge drinking 6, 268
biodynamic viticulture 19
biogenic amines 7
bisulphite compound 59, 142, **171**
blending 145
blind tasting 262
blue fining 153
Botrytis cinerea 142, 180
bound sulphur dioxide 172
bouquet 136
BRC 282
Brettanomyces 73, 156, **257**
British wine 71
Brix 231
bronze 34, 154, 176
Bual 124
buffering 25, **238**
Business Excellence 283

C

calcium 240
calcium carbonate 64
calcium phytate 155
calcium sulphate 63
calcium tartrate 63, 64
calcium tartrate-malate 65
cancer 5
Candida 72
canopy 13
capsules 209
carbonation 117
carbon cycle 3, 74
carbon dioxide 2, **34**, 44, 71, 74, 82, 85, 88, 92, 94, 95, 98, 107, 117, 145, 213, 255
carbon footprint 214
carbonic maceration 41, **97**
carbonyl compounds 72
carboxymethylcellulose 163
cardboard brick **201**, 216
cartridge filter 189
cascade 38
casein 151
cask wines 199
casse 245, **256**
catalyst 4, 34, 154, 170

caustic soda 238
CCP 271
cellulose 187
centrifuge **60**, 82, 144
cerasuolo 101
chalk 64
champagne 50
chapeau 90
chaptalisation 65
Charmat 116
cherry 126
chestnut 126
chiaretto 101
Chitin-glucan 156
chitosan 156
chlorinated sterilants 254
chlorine 203, 216
chlorophyll 2
cholesterol 6
chromatogram 239
citric acid 63, **176**, 243
clarete 101
clarification 60
clone 10
closures 202
cloudiness 241
Codex Alimentarius 270
col de poisson 151
cold soak 90
cold stabilisation 158
cold sterile filtration 223
colloids 147, 159, 240
colmated cork 203
comparative tasting 262
complexity 38
concrete vats 211
conductivity 240
contact process 159
continuous screw press 55
cool fermentation 103
copper 24, 34, 153, 177, 245
copper casse 256
copper salts 239
copper sulphate 176, 245, 250
copper sulphide 245
cream of tartar 252

criadera	121
cross-flow filter	192
crusher	47
cryo-concentration	68
cryo-extraction	68
cultured yeasts	77
cuve close	116
cuvée non-filtre	215
cyanide	4, 154
cyanidin	26

D

deacidification	**64**, 85
débourbage	60
decanting	260
délestage	94
delphinidin	26
denaturing	147
densimeter	231
density	76, **231**, 239
depth filtration	183
de-stemming	46
diacetyl	86
Diam	204
diammonium phosphate	69
diatomaceous earth	184
diatoms	184
dissolved oxygen	**36**, 53, 74
DMDC	217
DNA analysis	246
dosage	116
double pasta	102
double-salt	65
drainage	15
drainer	48
dry ice	34
Ducellier tanks	95
due diligence	290

E

earth filter	185
egg white	150
égrappoir	45
Eiswein	110

electrodialysis	161
ellagitannins	128
enrichment	66
enzymes	**4**, 71, 170
epoxy resin	88
erythorbic acid	174
esterification	139
esters	28, 136, 226, 238
estufagem	124
ethanol	75, **135**, 142, 233
ethyl acetate	255
European regulations	58
extracellular	98

F

Fagaceae	126
Fehling's solution	141, 239
fermentation	**71**, 95
fermentation vessels	88
ferrocyanide	154
filling level	220
filtration	82, 148, **181**, 214
fining	**146**, 214
fino	122, 142
flame photometer	246
flash détente	97
flash pasteurisation	222
flavonoids	6
flavour components	**27**, 37
flavour scalping	201, 206
flex cracking	200
flexitanks	214
flor	72, **121**, 123
flotation	39, **61**
flow diagram	272
fluorescent lights	196
foreign bodies	254
fortification	82
fortified wines	117
fouloir	47
free-run juice	47
free sulphur dioxide	**171**, 241, 250
French Paradox	5
fructose	2, 22, 66
fruit wine	71

fungi 72
fusel oils 136

G

galets 13
gamma radiation 223
Ganimede tank 89
gas chromatography 234
gas flushing 54
GC/MS/MS 246
gelatine 150
genetic modification 11
geranium taint 175, 258
glass 254
glass bottle 195
gluconeogenesis 23
glucose 2, 22, 66
glycerine 141
glycerol 28, 72, **141**, 235
gönci 128
grafting 12
grape tannins 70
green harvest 17
gross lees 144
guard filter 191
gum arabic 147, 177
gypsum 63
gyropallets 115

H

HACCP 214, 254, **270**, 290
haemocytometer 81
hand harvesting 40
Hansenula 72
haze 241
HDPE 200
headspace 212
heart disease 5
helium 34
hemi-cellulose 130
herbaceous plants 15
histamine 7
horizontal screw press 50
HPLC 239, 246

hydrocarbides 28
hydrogen peroxide 174, 216, 242
hydrogen sulphide 12, 31, 33, 69, 106
176, 198, 207
hydrolysis 226
hydrometer 231
hyperoxidation 38, 39, **62**
hypochlorite 216

I

ichthyocol 151
industrial wines 78
inert gas **33**, 212
ingredient listing 165
ingredients 149
intracellular fermentation 98
iodine 242
ion exchange 160
iron 24, 121, 153, 245
iron casse 256
irrigation 16
isinglass 151
ISO 14001:2004 280
ISO 22000:2005 280
ISO 9000 278

K

Kabinett 110
ketones **142**, 171, 226
kieselguhr **183**, 223
kieselsol 152
'killer' yeast 77
Kloeckera apiculata 72
KMW 231
krypton 34

L

laccase 180
lactic acid 84, 138
lactic bacteria 179
Lactobacillus 85
lactones 73
lagar 118

lead foil 209
lees 106, 116, 185
lenticels 203
Leuconostoc 85
liqueur de tirage 114
liqueur d'expedition 115, 213
liqueur wines 117
lot mark 288
Louis Pasteur 218
low density polyethylene 210
low intervention 143
lysozyme 179

M

maceration 45
macération carbonique 97
macération pelliculaire 104
machine harvesting 43
mad cow disease 150
maderised 250
magnesium 24
malic acid **23**, 64, 85, 169
Malmsey 124
malo-lactic fermentation **84**, 142
malvidin 26
mannoproteins 163
Marsala 124
maturation 225
maturation in bottle 226
MCB 196
mechanical lagar 119
membrane filter 189
metatartaric acid **162**, 243
methanol 136, 234
méthode champenoise 113
méthode traditionelle 113
microbiological analysis 247
microclimate 12
microorganisms 57, 71
micro-oxygenation 38, 120, 128, **133**
minerals 14, **24**, 60
mineral salts 3, 235
minimum intervention 164
MLF 84
modern bottling 215

molecular sulphur dioxide 238
molecular weights 75
montmorillonite 151
Moscato 117
Möslinger 153
mousiness 257
Muscadet 106
Muscat 118
muselet 116
must 39, 59
must concentration 67
musty taint 254

N

natural cork 202
natural fermentation 76
near infra-red 234
neon 34
NIR 234
nitrogen 3, 28, 31, 34, **35**, 61, 69, 107, 212
noble rot 110
nominal quantity 289
non-volatile acids 235

O

oak 126
oak barrels 118
oak chips 132
ochratoxin A 156
Oechsle 231
Oenococcus oeni 85
off-flavours 60
oloroso 122
One-Plus-One cork 204
organic viticulture 19
organic wine 19
organic winemaking 59
osmotic pressure 83
ox blood 149
oxidases 4, 34, **178**
oxidation 4, 6, 59, 81, 105, **107**, 154, 249
oxidative coupling 125
oxidising enzymes 43

oxygen	**31**, 61, 123, 125	polysaccharide	131, 177
	133, 200, 225	polyvinyl alcohol	200
oxygen barrier	205, 207	polyvinylpolypyrrolidone	152
oxygen permeability	195	pompe bicyclette	117
		population density	81
P		pore size	192, 224
		porous pot	133
pad filter	186	Port styles	120
partial root drying	16	post-fermentation maceration	90
Pasteur	1, 84	potassium	**7**, 240
pasteurisation	82	potassium bicarbonate	64
pathogenic bacteria	214	potassium bisulphate	63
pathogenic microorganisms	211	potassium bisulphite	167
PDO	287	potassium bitartrate	64, 159, **240**, 252
pectinolytic enzymes	48, **178**	potassium chloride	161
pectins	48	potassium ferrocyanide	153
Pediococcus	85	potassium iodate	242
Pedro Ximenez	123	potassium metabisulphite	**33**, 44, 166
penicillium mould	203	potassium sorbate	175
peonidin	26	pre-ferment maceration	90
peracetic acid	216, 254	premature fermentation	59
pergola	13	primary aroma	27
permitted additive	241	primary fermentation	84
peroxide	203	processing aids	149
peroxide method	242	proof spirit	232
PET bottles	197, 252	protective colloids	252
petri dish	248	protein stability	241
PGI	287	proteins	**28**, 69, 70, 143
pH	25, 173, **237**, 240	pumping-over	93
phenols	73	punching down	91
phosphomolybdic acid	241	puttonyos	112
photosynthesis	2, 21	PVA	200
phylloxera	9, **11**	PVC	210
physiological ripeness	29	PVC bottles	197
phytoalexins	7	PVI/PVP	155
Pichia	72	PVPP	152
pigeage	**90**, 120	pyknometer	231
pips	48, 94		
planting density	15	**Q**	
plastering	63		
plate and frame filter	186	quality assurance	269
pneumatic press	53	quality control	269
polylaminated capsule	210	quality plan	228
polyphenoloxidase	70, 97, 180	Quercus	126
polyphenols	5, 48, 49, 59, 87, 95, 104	Quercus alba	127
	119, 129, 132, 134, 235	Quercus mongolicus	127

Quercus petraea 127
Quercus robur 127
Quercus sessilis 127

R

rack and return 94
racking 144
rancio 118
RCGM 66, 140, 213
records 228
reducing sugars 141
reduction 62
reductive condition 106
reductive taint 33, 38, 176, **250**
re-fermentation 175, 215
refractive index 22, 239
refractometer 22
refrigeration 253
remuage 114
residual sugars **140**, 239
resveratrol 7
reverse osmosis 68
rinsing 219
rootstock 12
rosado 101, 102
rosato 101
rotary fermenters 96
rotary vacuum filter 184

S

Saccharomyces 72
Saccharomyces bayanus 72
Saccharomyces cerevisiae 77, 247
Saccharomycodes 72
saignée 102
salmonella 150
salt 161
Sauternes 83
screwcaps 38, **206**
seasoning 129
secondary fermentation 84
second fermentation 255
selective membranes 161
semi-macération carbonique 100

semi-permeable membrane 68, 83
Sercial 124
settling 60
sheet filter 186
shelf-life 63, 199
sherry method 121
silica sol 152
skimmed milk 151
skin contact 27, 79, 90, **104**
sodium 7, **246**
soil 15
solera 121
sorbic acid **175**, 243, **258**
sorting 45
sparging 35, **37**
Spätlese 110
specific gravity 231
spice 126
stable colloid 177
stainless steel 176
stalks 46
sterile bottling 215
sterile techniques 223
sterilisation 216
stuck fermentation 179
suberin 204
submerged cap 92
succinic acid 138
sucrose 2, 21, 66, 114, 140, 213
sugar 57
sugar-free extract 235
sugars 3
sulphur 165
sulphur dioxide 4, 36, 73, 76, 82, 107
139, 142, 145, **165**, 179, 198
201, 207, 213, 223, 256
sulphuric acid 166, 236, 238, 242
sulphuring 166
super-critical carbon dioxide 204
surface effect **212**, 226, 251
sur lattes 114
sur lie 106
sur points 115
Süssreserve **109**, 140, 213
sweet wines 108
sweetening **65**, 213

synthetic closures   205
Systematic Tasting   265

T

TA   237
Tabarié formula   235
tangential filter   192
tank method   116
tank press   53
tannins   **25**, 94, 125, 134, 151
tartaric acid   **23**, 63, 162, 240
tartrate crystals   143, 162, 252
tartrates   177
tartrate stability   240
tasting glass   261
tasting note   263
TCA   **203**, 216, 254
TDE   235
technical corks   203
teinturier   26
temperature   78, 259
terpenols   28
Terra Vitis   18
terre rose   186
terroir   **17**, 20
tertiary aromas   207
thermotic bottling   219
thermo-vinification   96
thiamine   70
tin capsule   209
tin-lead capsule   209
titratable acidity   235
titration   236
toasting   129
Tokaji Aszú   83, **111**
tolerable negative error   289
Torulopsis   72
total acidity   236
total dry extract   235
total sulphur dioxide   172, 242
TQM   270
traceability   18, 229, **288**
traditional bottling   214
traditional method   113
transfer method   116

triage   40, **45**
trichloranisole   **203**, 216, 254
Trockenbeerenauslese   110
tunnel pasteurisation   220
tyrosinase   180

U

ultra-filtration   194
unstable colloids   177
UV radiation   196, 226

V

vacuum distillation   68
vanilla   126
vanillin   130
varietal character   178
Vaslin   50
vegans   144, 149
vegetarians   144, 149
véraison   17, **29**
Verdelho   124
vertical screw press   49
vin clair   114
vin d'une nuit   102
vinegar   4, 49, 88, 138, 239, 255
Vino-lok   208
vins doux naturels   82, **118**
Vitaceae   9
vitamin C   33, **174**
vitamins   60
Viticulture Raisonée   18
Vitis   9
volatile acidity 81, 138, 179, 215, **238**, **255**
volatile acids   226

W

water   74
weather   23
whole bunch fermentation   100
whole bunches   105
wild ferments   77
wild yeasts   72
Willmes   53

wine growing zones   58
wine yeasts   72

X

xenon   34

Y

yeast   **72**, 169
yeast bitten   144
yeast nutrient   83

Z

zeta potential   188
Zork   208

The Wine Business Library

ORGANIC WINE: A MARKETER'S GUIDE
Béatrice Cointreau

Building on detailed case studies, Cointreau presents an exhaustive analysis of global production and market trends, and provides clear insights on how to position one's product to the best effect.

$29.95
ISBN 978-1935879633
Pub Date: October 1, 2015
Paperback, 6 x 9 inches, 200 pp., graphs and charts

WINE MARKETING & SALES, 2ND EDITION
Paul Wagner, Janeen Olsen, Ph.D., and Liz Thach, Ph.D, foreword by Robert Mondavi

This completely revised and updated edition of the bestselling book puts new, practical, and powerful strategies into the hands of veteran brand managers and marketing professionals, and the vast bank of wine marketing knowledge within reach of the nascent winery owner.

$75.00
ISBN 978-1-934259-25-2
Hardcover, 7 x 10 inches, 400 pp., illustrations and fully indexed

WINE BUSINESS CASE STUDIES: THIRTEEN CASES FROM THE REAL WORLD OF WINE BUSINESS MANAGEMENT
Pierre Mora, Editor

Published in association with the Bordeaux College of Business, this book applies business pedagogy's powerful learning tool to the unique challenges of wine business management. *Wine Business Case Studies* is written by an international group of respected wine business scholars.

$30.00
ISBN 978-1-935879-71-8
Paperback, 8.5 x 11 inches, 300 pp., graphs and charts

THE BUSINESS OF WINEMAKING
Jeffrey L. Lamy

Places all facets of the wine business in perspective for investors, owners, and anyone else who is interested in how the wine business operates.

$45.00
ISBN 978-1-935879-65-7
Paperback, 7 x 10 inches, 360 pp., 250 illustrations, charts, graphs, and fully indexed

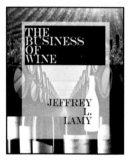

WINE MARKETING ONLINE
Brue McGechan

The whole wired realm of wine marketing is revealed in this encyclopedic yet readable and easy-to-follow guide.

$29.95
ISBN 978-1-935879-87-9
Paperback, 6 x 9 inches, 418 pp., illustrations and fully indexed

HOW TO LAUNCH YOUR WINE CAREER
Liz Thach, Ph.D. & Brian D'Emilio, foreword by Michael Mondavi

Career coaching from two of wine's most respected professionals and scores of industry icons like winemaker Heidi Barrett and writer James Laube of the *Wine Spectator*.

$29.95
ISBN 978-1-934259-06-1
Paperback, 6 x 9 inches, 354 pp., fully indexed